Maths.
Dept.

MATHEMATICAL A
AND TECHNIQ

OXFORD MATHEMATICAL HANDBOOKS

General Editor
JOHN CRANK

OXFORD MATHEMATICAL HANDBOOKS

Mathematical Analysis and Techniques

A. PAGE

VOLUME I

OXFORD UNIVERSITY PRESS
1974

Oxford University Press, Ely House, London W. 1

GLASGOW NEW YORK TORONTO MELBOURNE WELLINGTON
CAPE TOWN IBADAN NAIROBI DAR ES SALAAM LUSAKA ADDIS ABABA
DELHI BOMBAY CALCUTTA MADRAS KARACHI LAHORE DACCA
KUALA LUMPUR SINGAPORE HONG KONG TOKYO

ISBN 0 19 859612 ×

© OXFORD UNIVERSITY PRESS 1974

C.F. Mott College of Education

PRINTED IN NORTHERN IRELAND AT THE UNIVERSITIES PRESS, BELFAST

Preface

In writing this book I have had two classes of readers in mind. The rapid growth in technical education in the last decade has led to a demand for textbooks which cater for students who are concerned with mathematics as a tool of the sciences and whose main interest lies in the practice of mathematics rather than the refinements of its theories. Such students require a treatment of analysis which, while precise and rigorous in its formulations of theorems, is based on a somewhat broader framework of assumptions than would be appropriate for a student reading for an honours degree in mathematics. But I have also had in mind those students for whom mathematics is an absorbing intellectual study in its own right with a beauty unsurpassed in any other intellectual discipline, and I hope that the book will form a useful introduction to the more abstract university texts for honours students in mathematics, while recognizing that this category of student will require to go more deeply into some matters later on.

In the present ferment of ideas on the content and methods of mathematics courses in schools and universities, there has been a tendency on the part of some critics to denigrate the mastery of skills and techniques which are an essential part of traditional mathematics and prefer a more philosophical approach to the subject with emphasis on the exciting unifying concepts which have emerged over the last few years. There is no essential divergence in these two sets of aims. It is easy, of course, for the teaching of mathematics to degenerate into mere acquisition of techniques without reference to their place in the whole scheme of mathematical and scientific thinking, and this is no less true of modern than of traditional mathematics but I hope that I have kept a balance between ideas and techniques. For the student who is concerned with mathematics as a tool of the sciences, mastery of techniques has, if anything, been enhanced in importance by modern developments; for example, in the introduction of computer science, where the decision as to how to present information to the computer in the most economical form often demands a high degree of technical skill. Further, it is a cardinal principle of the author's philosophy in teaching mathematics that enjoyment in and a deep understanding of the fundamental ideas of the subject will arise not from a knowledge of the bare bones of abstract structures in the subject, but by doing mathematics in a wide range of applications of the structures.

The book should be useful for students reading for general degrees in universities and polytechnics and for those taking mathematics as an ancillary subject as part of a degree in science or engineering. It would also cater for students studying for Higher National Diplomas and for students offering mathematics as a special subject in Colleges of Education.

A characteristic feature of the book is the variety of problems set for practice by students. The numerous worked examples in the text make the book especially useful for those who are studying privately for degrees.

I have to thank the Syndics of the Cambridge University Press and the Senate of the University of London for kind permission to use questions from their examination papers. I should also like to take this opportunity of thanking the publishers from whom I have had excellent co-operation during the preparation of the book for printing, and also the publishers' readers and Professor Crank, the general editor of the series, for their helpful and constructive criticisms.

A. P.

1973

Contents

VOLUME I

1. **CONVERGENCE OF SERIES AND INTEGRALS** — 1
   - Convergence of series of positive terms — 1
   - Convergence of series of alternating positive and negative terms — 9
   - Convergence of series of terms of any sign — 10
   - Convergence of series of complex terms — 11
   - Integrals with infinite range — 14
   - Integrals with infinite integrands — 19
   - Estimating the value of a definite integral — 21
   - Convergence of infinite series and infinite integrals — 28
   - Sum of a deranged series — 32

2. **POWER SERIES** — 40
   - Differentiation of finite series — 40
   - Differentiation of infinite series — 41
   - Differentiation of power series — 41
   - Integration of power series — 43
   - Leibniz's theorem — 47
   - Maclaurin's theorem — 50

3. **THE MEAN VALUE THEOREM AND ITS APPLICATIONS** — 58
   - Rolle's theorem — 58
   - The first mean value theorem — 63
   - The general mean value theorem — 66
   - Applications of Taylor's theorem — 72
   - Further applications of Taylor's theorem — 76

4. **TECHNIQUES OF INTEGRATION** — 87
   - Techniques of integration — 87
   - Simply- and multiply-connected regions — 91
   - Asymptotes of a curve — 94
   - Equations of curves — 97
   - Polar equations and tangential–polar equation — 98
   - Some interesting curves — 105

CONTENTS

5. **FUNCTIONS OF MORE THAN ONE VARIABLE** — 118

   Analytical definition of $\frac{\partial z}{\partial x}, \frac{\partial z}{\partial y}$ — 120
   Maximum and minimum values of a function of two variables — 124
   Line of best fit — 125
   Technique of partial differentiation — 141
   Laplace's equation — 148

6. **MAXIMUM AND MINIMUM VALUES OF A FUNCTION OF TWO VARIABLES** — 157

   Further topics in partical differentiation — 157
   Taylor's theorem for a function of two variables — 157
   Maximum and minimum values of $V = f(x, y, z)$ when there are constraints on the variables x, y, z — 162
   Maximum and minimum values by Lagrange's method of undetermined multipliers — 164
   Directional derivatives — 166
   Differentiations of integrals containing a parameter — 168

7. **DOUBLE INTEGRALS** — 173

   Area of a closed curve — 175
   Line integrals — 187
   Double integrals — 190
   Green's theorem — 202
   Change of variables in a double integral — 206

8. **THE SYMBOLISM OF MODERN ALGEBRA** — 222

   Sets — 222
   The algebra of sets — 227
   Groups — 231

   ANSWERS TO EXERCISES — 237

   INDEX — 251

VOLUME II

9. **DIFFERENTIAL EQUATIONS** — 1
10. **PARTIAL DIFFERENTIAL EQUATIONS** — 65
11. **FOURIER SERIES** — 82
12. **MATRICES** — 110
13. **VECTOR FIELDS** — 164
14. **FUNCTIONS OF A COMPLEX VARIABLE** — 189
15. **INTEGRATION** — 219
16. **EVALUATION OF INTEGRALS** — 250

1
Convergence of Series and Integrals

We begin this chapter with a systematic study of the concept of convergence of an infinite series. Consider the series of *real* terms $\sum_{n=1}^{\infty} u_n$, sometimes abbreviated to $\sum u_n$, and let us denote the sum of the first n terms of the series by S_n, so

$$S_n = u_1 + u_2 + \cdots u_n.$$

If S_n tends to a unique limit S as n tends to infinity, the infinite series is called convergent and we call S the sum of the series; if the limit of S_n does not exist, we shall call the series divergent.

If the series $\sum u_n$ converges, then

$$\lim_{n \to \infty} u_n = 0.$$

For
$$\lim_{n \to \infty} u_n = \lim_{n \to \infty} (S_n - S_{n-1})$$
$$= \lim_{n \to \infty} S_n - \lim_{n \to \infty} S_{n-1} = S - S = 0.$$

The converse of this theorem is not true. For example, the series

$$1 + \tfrac{1}{2} + \tfrac{1}{3} + \tfrac{1}{4} + \cdots$$

diverges, although u_n tends to 0.

Convergence of series of positive terms

The theory of the convergence of series of positive terms is relatively simple and we consider such series first.

(i) *Partial sums with an upper bound*

A series of positive terms will converge if S_n is bounded above, i.e. if $S_n < K$, where K is a finite constant, for all values of n. For then, since the sequence of partial sums $S_1, S_2, S_3, \ldots$ is steadily increasing,

S_n must either tend to a limit or tend to infinity; the latter alternative is ruled out by the condition $S_n < K$.

Example. For the series $\sum_{n=1}^{\infty} \frac{1}{n^2}$,

$$S_n = \frac{1}{1^2} + \frac{1}{2^2} + \frac{1}{3^2} + \cdots + \frac{1}{n^2}$$

$$< 1 + \frac{1}{1.2} + \frac{1}{2.3} + \cdots + \frac{1}{(n-1)n}$$

$$= 1 + (1 - \tfrac{1}{2}) + (\tfrac{1}{2} - \tfrac{1}{3}) + \cdots + \left(\frac{1}{n-1} - \frac{1}{n}\right)$$

$$= 2 - \frac{1}{n} < 2.$$

Hence the series converges.

(ii) *The comparison theorem*

If $u_n > 0$, $v_n > 0$, and $v_n \leq ku_n$ for all values of n, then if Σu_n converges, so does Σv_n. For

$$v_1 + v_2 + \cdots + v_n \leq k[u_1 + u_2 + \cdots + u_n] < kS,$$

where S denotes the sum of the series Σu_n. The result follows.

If $u_n > 0$, $v_n > 0$, and $v_n \geq ku_n$ for all values of n, then if Σu_n diverges, so does Σv_n. For

$$v_1 + v_2 + \cdots + v_n \geq k(u_1 + u_2 + \cdots + u_n)$$

$$= kS_n.$$

Now $S_n \to \infty$ as $n \to \infty$ because Σu_n diverges, so $kS_n \to \infty$, hence Σv_n diverges.

Example. Consider the series Σu_n, where $u_n = \frac{n+3}{n^4+2}$. For large values of n, u_n approximates to $\frac{n}{n^4} = \frac{1}{n^3} < \frac{1}{n^2}$, which since $\Sigma \frac{1}{n^2}$ converges, suggests that Σu_n converges. To formalize the argument, $u_n < \frac{4n}{n^4} = \frac{4}{n^3} \leq \frac{4}{n^2}$, but $\Sigma \frac{4}{n^2}$ converges, hence Σu_n converges.

Example. Consider the series Σu_n, where $u_n = \dfrac{n+2}{2n^2+5}$. For large values of n, u_n approximates to $\dfrac{n}{2n^2} = \dfrac{1}{2n}$. Now $\Sigma \dfrac{1}{2n}$ diverges, which suggests that Σu_n diverges. To formalize the argument, $u_n > \dfrac{n}{2n^2+5n^2} = \dfrac{1}{7n}$, but $\Sigma \dfrac{1}{7n}$ diverges, hence Σu_n diverges.

(iii) *The ratio theorem*

If $u_n > 0$, $v_n > 0$ for all values of n, and if $\dfrac{v_n}{u_n}$ tends to a finite non-zero limit, λ, as n tends to infinity, then if Σu_n converges, so does Σv_n. If Σu_n diverges, so does Σv_n.

For the existence of the limit implies that, given any small positive ε, N exists such that v_n/u_n lies between $\lambda - \varepsilon$ and $\lambda + \varepsilon$ for all values of $n > N$. Hence, by taking $\varepsilon = \frac{1}{2}\lambda$, v_n/u_n lies between $\frac{1}{2}\lambda$ and $\frac{3}{2}\lambda$ if $n > N$. We can disregard the first N terms which do not affect the convergence or divergence of the series. From the fact that $v_n < \frac{3}{2}\lambda u_n$, we deduce that if Σu_n converges, so does Σv_n. And from the fact that $v_n > \frac{1}{2}\lambda u_n$, we deduce that if Σu_n diverges, so does Σv_n.

Example. If Σu_n is a convergent series of positive terms with sum S, prove that the series $\Sigma \{u_n(u_1+u_2+\cdots+u_n)\}$ converges.

$$\lim_{n \to \infty} \frac{\{u_n(u_1+u_2+\cdots+u_n)\}}{u_n} = \lim_{n \to \infty} S_n = S.$$

The result follows from (iii).

Example. Prove that the series $\Sigma \dfrac{n}{4n^2+1}$ diverges.

$$\lim_{n \to \infty} \frac{\left(\dfrac{n}{4n^2+1}\right)}{\dfrac{1}{n}} = \lim_{n \to \infty} \frac{n^2}{4n^2+1} = \frac{1}{4},$$

and since the series $\Sigma \dfrac{1}{n}$ diverges, the given series diverges.

In estimating the value of u_n in order to construct a suitable comparison series for the investigation of the convergence or divergence of the series Σu_n, the following notation is useful.

Definition of order

If a_n and b_n are positive functions of n such that $Kb_n < a_n < K_1 b_n$ for all values of n, where K, K_1 are finite constants, we say that a_n is of the order of b_n and write $a_n = O(b_n)$. It follows that if $u_n > 0$, $v_n > 0$, $v_n = O(u_n)$, and Σu_n converges, then Σv_n converges. For example,

$$v_n = e^{1/n} - 1 - \frac{1}{n} = O\left(\frac{1}{n^2}\right),$$

$$p_n = 1 - \cos\left(\frac{\pi}{2n}\right) = O\left(\frac{1}{n^2}\right),$$

$$q_n = \log\left(\frac{n+1}{n-1}\right) - \frac{2}{n} = O\left(\frac{1}{n^3}\right).$$

It follows that the series Σv_n, Σp_n and Σq_n are convergent series.

(iv) D'Alembert's test

If $u_n > 0$, the series Σu_n will converge if $\lim\limits_{n \to \infty} \frac{u_{n+1}}{u_n} < 1$, and will diverge if $\lim\limits_{n \to \infty} \frac{u_{n+1}}{u_n} > 1$. If $\lim\limits_{n \to \infty} \frac{u_{n+1}}{u_n} = 1$, the test gives no information, the series may converge or diverge.

For, let $\lim\limits_{n \to \infty} \frac{u_{n+1}}{u_n} = \lambda < 1$, and let t denote the number midway between λ and 1. Since $\frac{u_{n+1}}{u_n} \to \lambda$, the ratio will be less than t if we take n to be large enough; it follows that there is a number p for which $u_{n+1} < tu_n$ whenever $n > p$. Denote the sum of the first p terms of the series by k_1 and the $(p+1)$th term by k. Then $u_{p+2} < kt$; $u_{p+3} < tu_{p+2} < kt^2$, etc. Thus,

$$S_n < k_1 + k(1 + t + t^2 + \cdots t^{n-p-1}) = k_1 + k\frac{(1 - t^{n-p})}{1 - t}$$

$$\therefore S_n < k_1 + \frac{k}{1-t} = K, \text{ say}.$$

Hence, $S_n < K$ and the series is convergent.

Example. If $u_n = \frac{3n+1}{2^n}$, $\frac{u_{n+1}}{u_n} = \frac{3n+4}{3n+1} \cdot \frac{1}{2}$, and $\lim\limits_{n \to \infty} \frac{u_{n+1}}{u_n} = \frac{1}{2}$.

Hence, the series Σu_n converges.

If $\lim_{n\to\infty} \frac{u_{n+1}}{u_n} > 1$, then for some q, $\frac{u_{n+1}}{u_n} > 1$ if $n > q$; thus u_n does not tend to zero as n tends to infinity and the series must diverge.

If $\lim_{n\to\infty} \frac{u_{n+1}}{u_n} = 1$, the series may converge or diverge. For example, $\lim_{n\to\infty} \frac{u_{n+1}}{u_n} = 1$ for the series $\Sigma \frac{1}{n^2}$, which converges; $\lim_{n\to\infty} \frac{u_{n+1}}{u_n} = 1$ for the series $\Sigma \frac{1}{n}$ which diverges.

Note that D'Alembert's test postulates the existence of a limiting value for u_{n+1}/u_n as n tends to infinity. This limit does not exist for all series. For example, for the convergent series Σu_n, where $u_n = 2^{-n}$ if n is odd and 3^{-n} if n is even, u_{n+1}/u_n tends to 0 if n is odd, and tends to infinity if n is even. Hence $\lim_{n\to\infty} u_{n+1}/u_n$ does not exist for this series.

(v) *Cauchy's test*

If $u_n > 0$, the series Σu_n will converge if $\lim_{n\to\infty} u_n^{1/n} < 1$. For then, if t denotes the number midway between the value of the limit and 1,

$$u_n^{1/n} < t$$

for all $n > q$, i.e. $u_n < t^n$. Hence, if k_1 denotes the sum of the first q terms of the series

$$S_n < k_1 + t^{q+1} + t^{q+2} + \cdots$$

$$< k_1 + \frac{t^{q+1}}{1-t} < k_1 + \frac{1}{1-t} = k_2, \text{ say.}$$

Hence the series Σu_n converges.

Example. The series Σu_n, where $u_n = \left(1+\frac{1}{n}\right)^{-n^2}$ converges since

$$\lim_{n\to\infty} u_n^{1/n} = \lim_{n\to\infty} \left(1+\frac{1}{n}\right)^{-n} = e^{-1} < 1.$$

We have seen that if $\lim_{n\to\infty} \frac{u_{n+1}}{u_n} = 1$ D'Alembert's test yields no information about the convergence or divergence of the series Σu_n. A useful test which covers some cases of this type is Kummer's test.

(vi) Kummer's test

The series Σu_n, $(u_n > 0)$, converges if a positive sequence ϕ_n exists such that

$$\lim_{n \to \infty} \left(\phi_n \frac{u_n}{u_{n+1}} - \phi_{n+1} \right) > 0.$$

This implies that a positive t exists such that,

$$\phi_n \frac{u_n}{u_{n+1}} - \phi_{n+1} > t > 0 \quad \text{for} \quad n \geq N.$$

It follows that $\phi_n u_n - \phi_{n+1} u_{n+1} > t u_{n+1}$. On writing down a set of inequalities for values of n from N to $n-1$, we see that

$$\phi_N u_N - \phi_n u_n > t(u_{N+1} + u_{N+2} + \cdots + u_n)$$

so

$$\phi_N u_N > t(S_n - S_N),$$

and it follows that

$$S_n < S_N + \frac{1}{t} \phi_N u_N = K, \text{ say},$$

$$S_n < K,$$

and hence the series converges.

The series Σu_n, $(u_n > 0)$, diverges if a positive sequence ϕ_n exists such that

$$\lim_{n \to \infty} \left(\phi_n \frac{u_n}{u_{n+1}} - \phi_{n+1} \right) < 0$$

and $\Sigma \dfrac{1}{\phi_n}$ diverges.

For then $\phi_n \dfrac{u_n}{u_{n+1}} - \phi_{n+1} < 0$ for $n \geq N$, whence

$$\phi_n u_n - \phi_{n+1} u_{n+1} < 0.$$

It follows that $\phi_N u_N < \phi_{N+1} u_{N+1} < \phi_{N+2} u_{N+2} \cdots < \phi_n u_n$. Hence

$$u_N + u_{N+1} + \cdots + u_n > u_N + \frac{\phi_N}{\phi_{N+1}} u_N + \frac{\phi_N}{\phi_{N+2}} u_N + \cdots$$

$$= u_N \phi_N \left(\frac{1}{\phi_N} + \frac{1}{\phi_{N+1}} + \frac{1}{\phi_{N+2}} + \cdots + \frac{1}{\phi_n} \right),$$

which tends to ∞ as n tends to ∞. Hence $S_n \to \infty$ and the series $\Sigma\, u_n$ diverges.

Kummer's test is a sensitive one; we shall consider its application in the case of a series $\Sigma\, u_n$ for which

$$\frac{u_n}{u_{n+1}} = 1 + \frac{\lambda}{n} + O\left(\frac{1}{n^2}\right).$$

First consider $\phi_n = n$ (it is well known that the series $\Sigma\, \dfrac{1}{n}$ diverges). Then the series $\Sigma\, u_n$ will converge or diverge according as

$$\lim_{n \to \infty} \left\{ n \frac{u_n}{u_{n+1}} - (n+1) \right\} > 0 \quad \text{or} \quad < 0,$$

i.e. $$\lim_{n \to \infty} \left\{ n\left(1 + \frac{\lambda}{n} + O\left(\frac{1}{n^2}\right)\right) - (n+1) \right\} > 0 \quad \text{or} \quad < 0,$$

i.e. $\lambda - 1 > 0$ or $\lambda - 1 < 0$. Hence the series converges if $\lambda > 1$ and diverges if $\lambda < 1$. Next, suppose $\lambda = 1$, and apply Kummer's test with $\phi_n = n \log n$ (it is well known that the series $\Sigma\, \dfrac{1}{n \log n}$ diverges). Then the series Σu_n will converge or diverge according as

$$\lim_{n \to \infty} \left\{ n (\log n) \frac{u_n}{u_{n+1}} - (n+1) \log(n+1) \right\} > 0 \quad \text{or} \quad < 0,$$

i.e. $$\lim_{n \to \infty} \left\{ n \log n \left(1 + \frac{1}{n} + O\left(\frac{1}{n^2}\right)\right) - \right.$$
$$\left. - (n+1)\left(\log n + \frac{1}{n} + O\left(\frac{1}{n^2}\right)\right) \right\} > 0 \quad \text{or} \quad < 0,$$

i.e. $-1 > 0$ or $-1 < 0$. It follows that the series diverges. Thus the series $\Sigma\, u_n$, for which $\dfrac{u_n}{u_{n+1}} = 1 + \dfrac{\lambda}{n} + O\left(\dfrac{1}{n^2}\right)$ converges if $\lambda > 1$ and diverges if $\lambda \leq 1$.

Example. Prove that the series Σu_n, where $u_n = \dfrac{1.3.5 \cdots (2n-1)}{2.4.6 \cdots (2n)} \dfrac{1}{\sqrt{n}}$, diverges.

$$\frac{u_n}{u_{n+1}} = \frac{2n+2}{2n+1} \cdot \frac{\sqrt{(n+1)}}{\sqrt{n}} = \left(1+\frac{1}{n}\right)\left(1+\frac{1}{2n}\right)^{-1}\left(1+\frac{1}{n}\right)^{\frac{1}{2}}$$

$$= \left(1+\frac{1}{n}\right)\left(1-\frac{1}{2n}+O\left(\frac{1}{n^2}\right)\right)\left(1+\frac{1}{2n}+O\left(\frac{1}{n^2}\right)\right)$$

$$= 1+\frac{1}{n}+O\left(\frac{1}{n^2}\right).$$

Hence the series diverges.

This can be proved directly by noting that

$$\tfrac{3}{2} > \tfrac{4}{3}, \tfrac{5}{4} > \tfrac{6}{5}, \ldots \frac{2n-1}{2n-2} > \frac{2n}{2n-1},$$

from which it follows that if $v = \dfrac{1.3.5 \cdots (2n-1)}{2.4.6 \cdots (2n)}$, $v > \dfrac{1}{(4n)v}$, i.e. $v > \dfrac{1}{\sqrt{(4n)}}$. It follows that $u_n > \dfrac{1}{2n}$ and hence the series Σu_n diverges.

Example. For the series Σu_n, where $u_n = \left\{\dfrac{1.3.5 \cdots (2n-1)}{2.4.6 \cdots (2n)}\right\} \dfrac{1}{2n+1}$,

$$\frac{u_n}{u_{n+1}} = \frac{(2n+2)(2n+3)}{(2n+1)^2} = 1+\frac{2}{n}+O\left(\frac{1}{n^2}\right).$$

Hence the series converges.

Exercises 1(a)

1. Discuss the convergence or divergence of the series Σu_n, where u_n has the following values:

$$\frac{1}{(2n-1)^2}; \quad \frac{n}{n^2+1}; \quad \frac{2^n}{n!}; \quad \frac{2n^2}{(n+1)^4};$$

$$\frac{n!}{n^n}; \quad \frac{1}{\sqrt{\{n(n+2)\}}}; \quad \frac{n+\sqrt{n}}{n^2+n+1}; \quad \frac{(\log n)^2}{n^3}.$$

CONVERGENCE OF SERIES AND INTEGRALS

2. x is positive; find the range of values of x for which the series Σu_n converges, where u_n has the following values:

$$\frac{x^n}{2n+1}; \quad \frac{x^{2n}}{(2n)!}; \quad \frac{n^2 x^{n+1}}{n+2}; \quad \frac{x^n}{1+x^{2n}};$$

$$\frac{(n!)^2 x^n}{(3n)!}; \quad \left(1-\cos\frac{\pi}{2n}\right)x^n.$$

3. Prove that the series $x+y^2+x^3+y^4+\ldots$, where $0 < x < y < 1$, converges.

4. In the series Σu_n, where $u_n = 1 - n \log \frac{2n+1}{2n-1}$, prove that $u_n = O\left(\frac{1}{n^2}\right)$ and hence that the series converges.

5. $a_n > 0$ and Σa_n converges. Prove that the series Σa_n^2 and Σa_n^3 converge.

6. $a_n > 0$, $b_n > 0$, and $\frac{a_n}{b_n}$ tends to a finite non-zero limit as n tends to infinity. Prove that if the series Σb_n converges, so do the following series:

$$\Sigma a_n; \quad \Sigma a_n b_n; \quad \Sigma a_n^2.$$

Convergence of series of alternating positive and negative terms

If $a_n > 0$ for all values of n, the series $a_1 - a_2 + a_3 \cdots$ will converge if the sequence $a_1, a_2, \ldots$ is steadily decreasing and $a_n \to 0$ as $n \to \infty$.

For then, the sequence of partial sums $S_1, S_3, S_5, \ldots$ is steadily decreasing because $S_{2n-1} - S_{2n+1} = a_{2n} - a_{2n+1}$, which is positive. Also

$$S_{2n-1} = (a_1 - a_2) + (a_3 - a_4) + \cdots + (a_{2n-3} - a_{2n-2}) + a_{2n-1}$$
$$> (a_1 - a_2).$$

Thus the sequence $S_1, S_3, S_5 \cdots$ is steadily decreasing and bounded below and, therefore, tends to a limit as n tends to infinity, say $\lim_{n\to\infty} S_{2n-1} = T_1$.

The sequence of partial sums $S_2, S_4, S_6, \ldots$ is steadily increasing because $S_{2n} - S_{2n+2} = -a_{2n+1} + a_{2n+2}$, which is negative. Also

$$S_{2n} = a_1 - (a_2 - a_3) - \cdots - (a_{2n-2} - a_{2n-1}) - a_{2n} < a_1.$$

Thus, the sequence $S_2, S_4, S_6 \cdots$ is steadily increasing and bounded above, and therefore tends to a limit as n tends to infinity, say

$\lim_{n\to\infty} S_{2n} = T_2$. Finally,
$$\lim_{n\to\infty}(S_{2n-1}-S_{2n}) = \lim_{n\to\infty} a_{2n} = 0.$$
$$\therefore T_1 - T_2 = 0 \quad \text{or} \quad T_1 = T_2.$$

It follows that S_n tends to a unique limit as n tends to infinity and hence the series converges.

Example. The series $\Sigma \dfrac{(-1)^n}{n^{\frac{1}{4}}}$ converges since $\dfrac{1}{n^{\frac{1}{4}}}$ steadily decreases as n increases and $\lim_{n\to\infty} \dfrac{1}{n^{\frac{1}{4}}} = 0$.

Convergence of series of terms of any sign

If the series $\Sigma |u_n|$ converges, then the series Σu_n converges. For if $a_n = \frac{1}{2}(|u_n|+u_n)$ and $b_n = \frac{1}{2}(|u_n|-u_n)$, Σa_n is a series of positive terms and $a_n \le |u_n|$, whence Σa_n converges. Similarly, Σb_n converges. It follows from the fundamental definition of convergent series that $\Sigma (a_n - b_n)$ converges, i.e. Σu_n converges.

If the series $\Sigma |u_n|$ converges, the series Σu_n is said to converge absolutely. The first line of attack in the investigation of a series of terms of any sign is to investigate whether the series is absolutely convergent, for which the whole battery of tests for convergence of series of positive terms can be used. In particular, by D'Alembert's test, the series Σu_n will converge absolutely if $\lim_{n\to\infty} \left|\dfrac{u_{n+1}}{u_n}\right| < 1$; the series will diverge if $\lim_{n\to\infty} \left|\dfrac{u_{n+1}}{u_n}\right| > 1$, for this condition implies that u_n does not tend to zero as n tends to infinity.

A series Σu_n may converge although the corresponding series of moduli $\Sigma |u_n|$ diverges, in this case we say that the series Σu_n converges conditionally. For example, the series
$$1 + \tfrac{1}{3} - \tfrac{1}{2} + \tfrac{1}{5} + \tfrac{1}{7} - \tfrac{1}{4} + \cdots$$
converges, but the corresponding series of moduli
$$1 + \tfrac{1}{3} + \tfrac{1}{2} + \tfrac{1}{5} + \tfrac{1}{7} + \tfrac{1}{4} + \cdots$$
diverges.

Example. For the series Σu_n, where $u_n = \dfrac{x^n}{1+2^n}$, $\lim_{n\to\infty} \left|\dfrac{u_{n+1}}{u_n}\right| = \tfrac{1}{2}|x|$.

CONVERGENCE OF SERIES AND INTEGRALS

It follows that the series converges absolutely if $|x| < 2$, and diverges if $|x| > 2$. If $x = 2$, $u_n \to 1$ as $n \to \infty$ and hence the series diverges; if $x = -2$, $u_n \to (-1)^n$ as $n \to \infty$, and hence the series diverges. Thus the series converges absolutely if $|x| < 2$, and diverges if $|x| \geq 2$.

Exercises 1(b)

1. Discuss the convergence of the series Σu_n, where u_n has the following values,
$$\frac{(-1)^n}{n^{\frac{1}{3}}} \,; \quad \frac{(-1)^n}{1+\sqrt{n}} \,; \quad \frac{(-1)^n}{\log(1+n)} \,; \quad (-1)^n \sin\frac{x}{n}.$$

2. Investigate the convergence or divergence of the series Σu_n where u_n has the following values,
$$\frac{x^n}{n!} \,; \quad \frac{t^n}{(2n+1)!} \,; \quad \frac{1+n}{1+n^2}x^{2n-1}; \quad \frac{n^3}{(n+1)!}y^{2n} \,; \quad \frac{t^n}{1+t^n} \,;$$
$$(-1)^{n-1}\{\sqrt{(n+1)} - \sqrt{n}\}; \quad \frac{(-y)^n}{n} \,; \quad (-1)^{n+1}\log\left(1+\frac{1}{n}\right).$$

3. Discuss the convergence or divergence of the series Σu_n, where u_n has the following values,
$$\sin\frac{x}{n^2} \,; \quad \frac{\sin^2 nx}{n} \,; \quad (2^n + 5^n)x^n; \quad \frac{x^n}{1+x^{2n}} \,;$$
$$(-1)^n \sin\left\{\frac{\pi(n+1)}{4n}\right\}.$$

4. If Σv_n is a convergent series of positive terms, prove that the series $\Sigma \dfrac{\sin v_n}{v_n}$ diverges.

5. Σa_n is a convergent series of positive terms; prove that the series $\Sigma a_n \cos nt$ and $\Sigma a_n \sin nt$ converge absolutely for all values of t. Prove also that the series $\Sigma a_n x^n$ converges absolutely if $|x| \leq 1$.

6. Prove that the series $\Sigma \dfrac{1.3.5.7 \cdots (2n-1)}{n!} 3^{-n}$ is convergent and that its sum is $\sqrt{3} - 1$.

7. If Σa_n is a divergent series of positive terms, prove that the series $\Sigma \dfrac{a_n}{1+a_n}$ also diverges, but that the series $\Sigma \dfrac{a_n}{1+n^2 a_n}$ converges.

Convergence of series of complex terms

The series Σw_n, where w_n is complex, is convergent if $w_1 + w_2 + \cdots + w_n$ tends to a finite limit as n tends to infinity. If $w_n = u_n + iv_n$, where

u_n and v_n are real, the series Σw_n will converge if and only if the series Σu_n and Σv_n converge. Now, since $|w_n| = \sqrt{(u_n^2+v_n^2)}$, $|u_n| \le |w_n|$ and $|v_n| \le |w_n|$. It follows that if the series $\Sigma |w_n|$ converges, the series Σu_n and Σv_n will converge absolutely, and hence the series $\Sigma (u_n+iv_n)$, i.e. the series Σw_n will converge.

Thus the series Σw_n will converge if the series $\Sigma |w_n|$ converges, and we then say that the series Σw_n converges absolutely.

Example. The series $1+\dfrac{z}{1!}+\dfrac{z^2}{2!}+\dfrac{z^3}{3!}+\cdots$ converges absolutely for all values of z by D'Alembert's test. The sum of the series is called $\exp(z)$, or e^z, and the series defines $\exp(z)$ for complex values of z.

Example. The series $z-\dfrac{z^3}{3!}+\dfrac{z^5}{5!}-\cdots$ converges absolutely for all values of z by D'Alembert's test. The sum of the series is called $\sin z$ and the series defines $\sin z$ for complex values of z.

Example. The series $1-\dfrac{z^2}{2!}+\dfrac{z^4}{4!}-\cdots$ converges absolutely for all values of z by D'Alembert's test. The sum of the series is called $\cos z$ and the series defines $\cos z$ for complex values of z.

Exercises 1(c)

1. Prove that $e^{iz} = \cos z + i \sin z$, $\sin z = \dfrac{e^{iz}-e^{-iz}}{2i}$ and $\cos z = \dfrac{e^{iz}+e^{-iz}}{2}$.

   If $z = re^{i\theta}$, prove that, as $r \to \infty$, $\tan z \to -i$ if $0 < \theta < \pi$; and $\tan z \to i$ if $-\pi < \theta < 0$.

2. Prove that the series $\Sigma \dfrac{z^n}{n^2}$ converges absolutely if $|z| \le 1$.

3. Find the condition for the convergence of the series
$$\sum_{n=1}^{\infty} \left(\dfrac{2z}{(2z+1)}\right)^{n-1}$$
and find its sum.

4. Prove that the series $\Sigma \dfrac{z^n}{1+z^{2n}}$ converges if $|z| < 1$, and if $|z| > 1$. Investigate the behaviour of the series on the circle $|z| = 1$.

5. Prove that if the power series $\Sigma a_n z^n$ converges for $z = z_1$, then the series converges absolutely for all values of z within the circle $|z| = |z_1|$.

6. Prove that if $|z| < 1$, the series
$$\sum \dfrac{2^{n-1}z^{2^{n-1}}}{1+z^{2^{n-1}}}$$

converges absolutely to a sum $\dfrac{z}{1-z}$ and investigate the behaviour of the series when $|z| \geq 1$.

Circle of convergence of a power series

Some power series converge for all values of z, e.g. $\Sigma \dfrac{z^n}{n!}$, and some converge for no values of z apart from an isolated value, e.g. $\Sigma (nz)^n$ which converges for $z = 0$ but for no other values of z. For all other power series $\sum_{n=0}^{\infty} a_n z^n$, a positive number R exists such that the series converges absolutely for all values of z satisfying $|z| < R$, and diverges for all values of z satisfying $|z| > R$. This value R is called the radius of convergence of the series, and the circle $|z| = R$ is called the circle of convergence. At a point on the circle of convergence the series may converge or diverge.

(i) For example the series $\sum_{n=0}^{\infty} (2z)^n$ converges absolutely if $|z| < \tfrac{1}{2}$ and diverges if $|z| > \tfrac{1}{2}$. The circle $|z| = \tfrac{1}{2}$ is the circle of convergence of the series.

(ii) The circle of convergence of the series $\sum_{n=1}^{\infty} \dfrac{z^n}{n}$ is $|z| = 1$.

Uniform convergence of series

It might be expected that for an absolutely convergent series $\Sigma u_n(z)$, where $u_n(z)$ is a continuous function of z within a region, the sum of the series $S(z)$ would also be a continuous function of z within the region. But the example $u_n = z^2(1+z^2)^{-n}$, where $S(z) = 1+z^2$ if $z \neq 0$, and $S = 0$ if $z = 0$, shows that $S(z)$ is not necessarily a continuous function of z. This phenomenon was first investigated in 1841 by Weierstrass who introduced the concept of uniform convergence to explain the result.

We say that the series $\Sigma u_n(z)$ converges to $S(z)$ if, given any positive ε, we can choose N such that $|S_n - S(z)| < \varepsilon$ if $n > N$. The value of N required will clearly depend on ε but if we can choose N dependent only on ε and independent of z for all z within a given region, then we will call the series uniformly convergent within that region. If R is the radius of convergence of the power series $\Sigma u_n z^n$ and $R_1 < R$, the series converges absolutely for $|z| \leq R_1$. Further, since for all values of z in this range, $|S - S_n|$ for $\Sigma u_n z^n$ is less than or equal to $|S - S_n|$ for $\Sigma |u_n R_1^n|$, it is clear that the series $\Sigma u_n z^n$

will converge uniformly for $|z| \leq R_1$. Thus all power series $\Sigma u_n z^n$ with radius of convergence R are uniformly convergent if $|z| \leq R_1$, where $R_1 < R$, and the sum of the series is a continuous function of z within the region.

Integrals with infinite range

We now consider what meaning can be attached to a definite integral in which one or both of the limits of integration are ∞ or $-\infty$.

Example. Consider $\int_1^\infty \dfrac{dx}{x^2}$.

$$\int_1^k \frac{dx}{x^2} = 1 - \frac{1}{k} \to 1 \quad \text{as} \quad k \to \infty.$$

We define $\int_1^\infty \dfrac{dx}{x^2}$ as $\lim\limits_{k \to \infty} \int_1^k \dfrac{dx}{x^2}$, whence $\int_1^\infty \dfrac{dx}{x^2} = 1$. We write the solution simply in the form

$$\int_1^\infty \frac{dx}{x^2} = \left[-\frac{1}{x} \right]_1^\infty = -0 + 1 = 1.$$

Example. Consider $\int_0^\infty \dfrac{dx}{1+x^2}$.

$$\int_0^k \frac{dx}{1+x^2} = \tan^{-1} k \to \tfrac{1}{2}\pi \quad \text{as} \quad k \to \infty.$$

We define $\int_0^\infty \dfrac{dx}{1+x^2}$ as $\lim\limits_{k \to \infty} \int_0^k \dfrac{dx}{1+x^2} = \tfrac{1}{2}\pi$. We write the solution simply in the form

$$\int_0^\infty \frac{dx}{1+x^2} = [\tan^{-1} x]_0^\infty = \tfrac{1}{2}\pi - 0 = \tfrac{1}{2}\pi.$$

Example. Consider $\int_2^\infty \dfrac{dx}{x}$.

$$\int_2^k \frac{dx}{x} = \log k - \log 2 \to \infty \quad \text{as} \quad k \to \infty.$$

No meaning is attached to $\int_2^\infty \dfrac{dx}{x}$.

In general, if $\int_a^k f(x)\,dx$ tends to a finite limit as k tends to ∞, we define

$$\int_a^\infty f(x)\,dx \quad \text{as} \quad \lim_{k\to\infty} \int_a^k f(x)\,dx,$$

and $\int_a^\infty f(x)\,dx$ is then called a convergent integral. If $\int_a^k f(x)\,dx$ does not tend to a finite limit as k tends to ∞, then no meaning is to be attached to $\int_a^\infty f(x)\,dx$, which is called a divergent integral.

Similar meanings are to be attached to integrals of the type $\int_{-\infty}^b f(x)\,dx$ and $\int_{-\infty}^\infty f(x)\,dx$.

Example. Consider the integral $\int_a^\infty \dfrac{dx}{x^n}$, where $a > 0$.

$$\int_a^k \frac{dx}{x^n} = -\left[\frac{1}{(n-1)x^{n-1}}\right]_a^k = -\frac{1}{n-1}\left(\frac{1}{k^{n-1}} - \frac{1}{a^{n-1}}\right) \quad \text{if} \quad n \neq 1,$$

and tends to a finite limit as $k \to \infty$ if $n > 1$, but not otherwise. Hence $\int_a^\infty \dfrac{dx}{x^n}$ converges if $n > 1$ but diverges if $n \leq 1$.

$$\int_a^\infty \frac{dx}{x^n} = \frac{1}{(n-1)a^{n-1}} \quad \text{if} \quad n > 1.$$

Example. Consider the integral $\int_1^\infty \dfrac{dx}{x(1+x)}$.

$$\int_1^k \frac{dx}{x(1+x)} = \int_1^k \left(\frac{1}{x} - \frac{1}{1+x}\right) dx$$

$$= [\log x - \log(1+x)]_1^k$$

$$= \left[\log\left(\frac{x}{1+x}\right)\right]_1^k = \log\left(\frac{k}{1+k}\right) - \log \tfrac{1}{2}.$$

As k tends to infinity, this expression tends to the value $-\log \tfrac{1}{2}$, or log 2. Thus,

$$\int_1^\infty \frac{dx}{x(1+x)} = \log 2.$$

We write the answer simply in the form

$$\int_1^\infty \frac{dx}{x(1+x)} = \left[\log\left(\frac{x}{x+1}\right)\right]_1^\infty = \log 1 - \log \tfrac{1}{2}$$

$$= \log 2.$$

But we cannot write $\int_1^\infty \dfrac{dx}{x(x+1)}$ as $\int_1^\infty \dfrac{dx}{x} - \int_1^\infty \dfrac{dx}{x+1}$, since neither of these integrals $\int_1^\infty \dfrac{dx}{x}$ or $\int_1^\infty \dfrac{dx}{x+1}$ converges.

As another example of a convergent integral, consider

$$\int_0^\infty \frac{2}{(x+1)(x+2)(x+3)} dx = [\log(x+1) - 2\log(x+2) + \log(x+3)]_0^\infty$$

$$= \left[\log \frac{(x+1)(x+3)}{(x+2)^2}\right]_0^\infty$$

$$= \lim_{x \to \infty} \log \frac{(x+1)(x+3)}{(x+2)^2} - \log \tfrac{3}{4}$$

$$= \log 1 - \log \tfrac{3}{4} = \log \tfrac{4}{3}.$$

Exercises 1(d)

1. Prove that $\int_{1}^{\infty} \dfrac{dx}{(x+1)(x+4)}$ is convergent.

Prove that
$$\int_{2}^{\infty} \frac{15x+3}{(x-1)(x+2)(2x+1)} dx = \log \frac{128}{5}.$$

2. Evaluate $\int_{0}^{\infty} \dfrac{dx}{(x^2+a^2)(x^2+b^2)}$; $\int_{0}^{\infty} \dfrac{x\,dx}{(x^2+a^2)(x^2+b^2)}$; $\int_{0}^{\infty} \dfrac{x^2\,dx}{(x^2+a^2)(x^2+b^2)}$, where a and b are positive.

3. Sketch the curve $y = \dfrac{1}{a^2+x^2}$, where $a > 0$, and find the volume of the solid formed by rotating the area between the curve and the x-axis through 4 right angles about the x-axis. Find also the position of the mean centre of the portion of the solid for which $x \geq 0$.

4. Prove that $\int_{0}^{\infty} \sin x \, dx$ and $\int_{0}^{\infty} \cos x \, dx$ are not convergent.

5. Evaluate the integrals
$$\int_{0}^{\infty} \frac{x\,dx}{(1+x^2)^3} \; ; \int_{0}^{\infty} \frac{x^2\,dx}{(1+x^2)^3} \; ; \int_{0}^{\infty} \frac{x^3\,dx}{(1+x^2)^3} \; ; \int_{0}^{\infty} \frac{x^4\,dx}{(1+x^2)^3}.$$

6. Sketch the curve $y = \dfrac{1}{(a^2+x^2)^3}$ and find the volume of the solid formed by rotating the area between the curve and the x-axis through 4 right angles about the x-axis.

7. Evaluate the integrals $\int_{1}^{\infty} \dfrac{dx}{x(x^5+1)}$; $\int_{1}^{\infty} \dfrac{dx}{x(x^6+1)}$; $\int_{1}^{\infty} \dfrac{dx}{x(x^{2n}+1)}$.

8. Evaluate $\int_{-\infty}^{\infty} \dfrac{dx}{1+x^2}$; $\int_{-\infty}^{2} \dfrac{dx}{4+x^2}$.

9. If $I_1 = \int_{0}^{\infty} f(x^2) \, dx$ is a convergent integral, prove that $I_2 = \int_{-\infty}^{\infty} f(x^2) \, dx$ is also convergent and that $I_2 = 2I_1$.

10. By splitting the range of integration into two parts $(0, 1)$ and $(1, \infty)$,

18   MATHEMATICAL ANALYSIS AND TECHNIQUES

and making a change of variables in one of these ranges, show that

$$\int_0^\infty \frac{\log x}{1+x^2}\,dx = 0.$$

11. Show that $\int_0^\infty \frac{dx}{x^2+x+1} = \frac{2\pi}{3\sqrt{3}}$, and evaluate $\int_0^\infty \frac{dx}{(x^2+x+1)^2}$.

12. If $\int_a^\infty f_1(x)\,dx$ and $\int_a^\infty f_2(x)\,dx$ are convergent integrals, prove that $\int_a^\infty \{f_1(x)+f_2(x)\}\,dx$ and $\int_a^\infty \{f_1(x)-f_2(x)\}\,dx$ are also convergent.

13. If $f(x) \geq 0$, $\phi(x) \geq 0$, and $\phi(x) \leq f(x)$ for all values of $x \geq a$, prove that if $\int_a^\infty f(x)\,dx$ converges, then $\int_a^k \phi(x)\,dx$ steadily increases as k increases and is bounded above; it follows that $\int_a^\infty \phi(x)\,dx$ is convergent.

14. If $|\phi(x)| \leq |f(x)|$ and $\int_a^\infty |f(x)|\,dx$ converges, prove that $\int_a^\infty |\phi(x)|\,dx$ also converges.

15. Let $f_1(x) = \tfrac{1}{2}(|\phi(x)|+\phi(x))$, $f_2(x) = \tfrac{1}{2}(|\phi(x)|-\phi(x))$. Prove that $0 \leq f_1(x) \leq |\phi(x)|$ and $0 \leq f_2(x) \leq |\phi(x)|$.

Note. It follows that if $\int_a^\infty |\phi(x)|\,dx$ converges, then $\int_a^\infty f_1(x)\,dx$ and $\int_a^\infty f_2(x)\,dx$ also converge. Hence by the result of (12), $\int_a^\infty \{f_1(x)-f_2(x)\}\,dx$ also converges, i.e. $\int_a^\infty \phi(x)\,dx$ converges.

16. By combining the results of (14) and (15) it follows that if

$$|\phi(x)| \leq |f(x)|$$

and $\int_a^\infty |f(x)|\,dx$ converges, then $\int_a^\infty \phi(x)\,dx$ also converges.

For example, $\int_1^\infty \frac{\sin x}{x^2}\,dx$ converges because $\left|\frac{\sin x}{x^2}\right| \leq \frac{1}{x^2}$, and $\int_1^\infty \frac{1}{x^2}\,dx$ converges.

17. Prove that $\int_1^\infty \frac{\sin x}{x}\,dx$ and $\int_1^\infty \frac{\cos x}{x}\,dx$ are convergent integrals.

18. Investigate the convergence or divergence of integrals $\int_1^\infty \frac{\sin x}{\sqrt{x}}\,dx$ and $\int_1^\infty \frac{\cos x}{\sqrt{x}}\,dx$.

Integrals with infinite integrands

Another class of integrals which requires careful investigation is that for which the integrand becomes infinite at a point within the range of integration. For example, consider the integral $I = \int_{-1}^{8} x^{-\frac{1}{3}} \, dx$, in which the integrand becomes infinite when $x = 0$. We cannot evaluate the integral as $[\frac{3}{2} x^{\frac{2}{3}}]_{-1}^{8}$ because the result

$$\frac{d}{dx}(\tfrac{3}{2} x^{\frac{2}{3}}) = x^{-\frac{1}{3}}$$

is not valid for $x = 0$, although it is valid for all other points of the range of integration.

We shall consider the integrals over the ranges $(-1, -h)$ and $(h, 8)$, where h is positive, and find the sum of the limiting values of these two integrals as h tends to 0.

$$\int_{-1}^{-h} x^{-\frac{1}{3}} \, dx = \tfrac{3}{2}[(-h)^{\frac{2}{3}} - (-1)^{\frac{2}{3}}],$$

and tends to the limit $-\tfrac{3}{2}$ as h tends to 0.

$$\int_{h}^{8} x^{-\frac{1}{3}} \, dx = \tfrac{3}{2}[8^{\frac{2}{3}} - h^{\frac{2}{3}}]$$

and tends to the limit 6 as h tends to 0.

We define the value of the integral $\int_{-1}^{8} x^{-\frac{1}{3}} \, dx$ as

$$\lim_{h \to +0} \int_{-1}^{-h} x^{-\frac{1}{3}} \, dx + \lim_{h \to +0} \int_{h}^{8} x^{-\frac{1}{3}} \, dx = \tfrac{9}{2}.$$

In general, if $f(x)$ becomes infinite for $x = c$, lying within the range $a \leq x \leq b$, we define the integral $\int_{a}^{b} f(x) \, dx$ as

$$\lim_{h \to +0} \int_{a}^{-h} f(x) \, dx + \lim_{h \to +0} \int_{c+h}^{b} f(x) \, dx,$$

provided the two limits exist. Such integrals are called convergent. If the two limits do not exist, but the limit of the sum of the two integrals does exist, i.e. if

$$\lim_{h \to +0}\left[\int_a^{c-h} f(x)\,dx + \int_{c+h}^b f(x)\,dx\right]$$

exists and is equal to L, say, then $\int_a^b f(x)\,dx$ is not convergent, but we call L Cauchy's principal value of the integral $\int_a^b f(x)\,dx$. For example, $\int_{-1}^2 \frac{dx}{x}$ is not convergent but it has a principal value log 2.

Example. Consider the integral $\int_{-1}^1 x^{-2}\,dx$. It is clear that $\int_{-1}^{-h} x^{-2}\,dx$ and $\int_h^1 x^{-2}\,dx$ each tend to infinity as h tends to 0. Thus the integral is not convergent.

Exercises 1(e)

1. Show that $\int_{-8}^1 \frac{1}{x}\,dx$ is not convergent but that $\int_{-8}^1 \frac{1}{x^{\frac{2}{3}}}\,dx = 9$.

2. Show that $\int_0^{33} \frac{1}{(x-1)^{\frac{5}{3}}}\,dx$ is not convergent but that $\int_0^{33} \frac{1}{(x-1)^{\frac{2}{5}}}\,dx = 15$.

3. Show that $\int_{-a}^b \frac{1}{x^p}\,dx$, where a and b are positive and p an integer, can be evaluated if $p < 1$, but cannot be evaluated if $p \geq 1$.

4. Show that if $a > 0$ the integral $\int_0^\infty \frac{dx}{(x+a)\sqrt{x}}$ is convergent, and find its value.

5. Prove that the integral

$$\int_{7a^3}^{16a^3} \frac{a\,dx}{x(x-8a^3)^{\frac{1}{3}}},$$

where $a > 0$, is convergent, and prove that its value is

$$\left[-\tfrac{1}{4}\log 28 + \frac{\sqrt{3}}{2}\tan^{-1}(3\sqrt{3})\right].$$

Estimating the value of a definite integral

If $b > a$, and $f(x) \geq 0$ throughout the range $a \leq x \leq b$, then

$$I = \int_a^b f(x)\,dx \geq 0.$$

In the trivial case, in which $f(x) = 0$ throughout the range of integration, $I = 0$. In all other cases, I represents the area bounded by the curve $y = f(x)$, the ordinates $x = a$, $x = b$, and the portion of the x-axis between $x = a$ and $x = b$, and is positive. For an alternative proof, we write

$$I = \int_a^b f(x)\,dx = [\phi(x)]_a^b = \phi(b) - \phi(a),$$

where $\phi'(x) = f(x)$. Since $f(x) \geq 0$, $\phi(x)$ must either be stationary, or increase, as x increases, whence, since $b > a$, $\phi(b) \geq \phi(a)$, i.e. $I \geq 0$.

We now remove the restriction that $f(x)$ is positive and consider any function of x for which we can find constants h and k such that

$$h \leq f(x) \leq k,$$

for all values of x in the range $a \leq x \leq b$. Then $f(x) - h \geq 0$ and $k - f(x) \geq 0$ throughout the range $a \leq x \leq b$.

Hence $\int_a^b \{f(x) - h\}\,dx \geq 0$ and $\int_a^b \{k - f(x)\}\,dx \geq 0$,

i.e.

$$\int_a^b f(x)\,dx - h(b-a) \geq 0, \qquad k(b-a) - \int_a^b f(x)\,dx \geq 0$$

so

$$h(b-a) \leq \int_a^b f(x)\,dx \leq k(b-a).$$

This is a useful theorem in estimating the value of a definite integral.

Example. The indefinite integral $I = \int (1+x^3)^{\frac{1}{2}}\,dx$ cannot be found simply but we can estimate the value of $I = \int_{\frac{1}{2}}^{1}(1+x^3)^{\frac{1}{2}}\,dx$ by noting that, in the range $\frac{1}{2} \leq x \leq 1$,

$$\sqrt{\tfrac{9}{8}} \leq (1+x^3)^{\frac{1}{2}} \leq \sqrt{2},$$

whence
$$\tfrac{1}{2}\sqrt{\tfrac{9}{8}} \leq \int_{\frac{1}{2}}^{1} (1+x^3)^{\frac{1}{2}}\,dx \leq \tfrac{1}{2}\sqrt{2}.$$

By dividing the range of integration into $(\tfrac{1}{2}, \tfrac{3}{4})$; $(\tfrac{3}{4}, 1)$ and estimating the corresponding integrals, we can show that

$$\tfrac{1}{4}(\sqrt{\tfrac{9}{8}}+\sqrt{\tfrac{91}{64}}) \leq \int_{\frac{1}{2}}^{1} (1+x^3)^{\frac{1}{2}}\,dx \leq \tfrac{1}{4}(\sqrt{\tfrac{91}{64}}+\sqrt{2}).$$

This gives $0.56 < I < 0.65$ and estimates I more accurately than the previous result, $0.53 < I < 0.71$. It will be clear that the process of subdividing the range of integration can be extended to give limits of increasing precision between which I must lie.

The trapezoidal rule and Simpson's rule provide more accurate approximations to the value of an integral and in a later chapter we shall estimate the errors associated with them and obtain inequalities of the type proved above.

Estimating the value of n! when n is large

In statistical problems one meets terms of the type $n!$, where n is a large integer. Without the use of a computer the calculation of the value of such terms is an arduous job. But the statistician may not be interested in the precise value of $n!$; he is content if he can estimate its value, and in doing this the theorem we have just proved is a useful tool.

It is clear that $\log x$ steadily increases as x increases, whence

$$\log 1 \;<\; \int_{1}^{2} \log x\,dx < \log 2,$$

$$\log 2 \;<\; \int_{2}^{3} \log x\,dx < \log 3,$$

.
.
.

$$\log(n-1) < \int_{n-1}^{n} \log x\,dx < \log n.$$

So that, if S_n denotes
$$\log 1 + \log 2 + \cdots + \log n = \log n!,$$
$$S_n - \log n < \int_1^n \log x \, dx < S_n,$$
i.e. $S_n - \log n < n \log n - n + 1 < S_n$, which gives
$$n \log n - n + 1 < S_n < (n+1) \log n - n + 1.$$
It follows that $\log n!$ lies between $n \log n - n + 1$ and
$$(n+1) \log n - n + 1;$$
and hence that $n!$ lies between $n^n e^{-n+1}$ and $(n^{n+1}) e^{-n+1}$. For example, 100! lies between $100^{100} e^{-99}$ and $100^{101} e^{-99}$. Now
$$e^{-99} = 10^{-99 \log e} = 10^{-99(0 \cdot 434\ 29)} \quad \text{and lies between } 10^{-43} \text{ and } 10^{-42}.$$
Hence 100! lies between $10^{200} \cdot 10^{-43}$ and $10^{202} \cdot 10^{-42}$, i.e. between 10^{157} and 10^{160}.

The inequality
$$n^n e^{-n+1} < n! < n^{n+1} e^{-n+1}$$
is somewhat crude and we shall now prove a more precise result.

Consider the area bounded by the curve $y = \log x$, the x-axis, and the ordinates $x = r$, $x = r+1$ (see Fig. 1.1). It is clear that this area

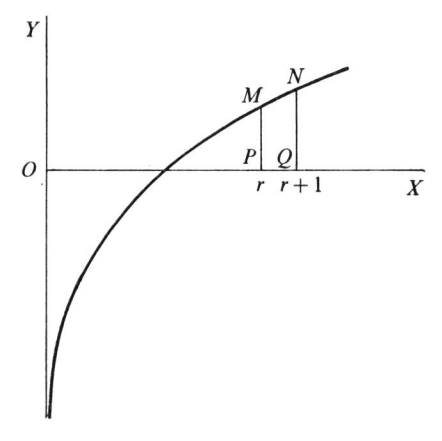

Fig 1.1

is greater than the area of the trapezium *PMNQ*. Thus
$$\int_r^{r+1} \log x \, dx > \tfrac{1}{2}(\log r + \log(r+1)).$$

Applying this inequality for $r = 1, 2, \ldots n-1$, we have
$$\int_1^n \log x \, dx > \tfrac{1}{2}\log 1 + \log 2 + \log 3 + \cdots + \log(n-1) + \tfrac{1}{2}\log n$$
$$= \log n! - \tfrac{1}{2}\log n.$$

Hence
$$n \log n - n + 1 > \log n! - \tfrac{1}{2}\log n,$$
i.e.
$$\log n! < (n+\tfrac{1}{2})\log n - n + 1,$$
$$\therefore n! < n^{n+\frac{1}{2}} e^{-n+1},$$

a result due to Stirling. We have now improved the upper bound of the previous inequality and can write
$$n^n e^{-n+1} < n! < n^{n+\frac{1}{2}} e^{-n+1}.$$

It is slightly more difficult to improve the lower bound of the inequality.

It may be proved that
$$n! = \sqrt{(2\pi)} \, n^{n+\frac{1}{2}} e^{-n} \left(1 + \frac{1}{12n} + \cdots\right).$$

Exercises 1(f)

1. Prove that, if $b > a$, and $f(x) \geq g(x)$ throughout the range $a \leq x \leq b$, then
$$\int_a^b f(x) \, dx \geq \int_a^b g(x) \, dx.$$

Prove that
$$\int_0^{\frac{\pi}{2}} \sin^p x \, dx \leq \int_0^{\frac{\pi}{2}} \sin^q x \, dx \quad \text{if} \quad p > q,$$
and that
$$\int_0^{\frac{\pi}{2}} \cos^p x \, dx \leq \int_0^{\frac{\pi}{2}} \cos^q x \, dx \quad \text{if} \quad p > q.$$

2. Prove that, if $b > a$, and $f(x) \leq 0$ throughout the range $a \leq x \leq b$, then
$$\int_a^b f(x) \, dx \leq 0.$$

CONVERGENCE OF SERIES AND INTEGRALS

3. If $b > a$, and m is the least value and M the greatest value of $f(x)$ in the range $a \leq x \leq b$, prove that

$$m(b-a) \leq \int_a^b f(x)\,dx \leq M(b-a).$$

4. Without evaluating the integrals, arrange the following integrals in order of increasing magnitude.

$$I_1 = \int_0^{\frac{\pi}{6}} \sin^5 x\,dx, \quad I_2 = \int_0^{\frac{\pi}{6}} \cos^5 x\,dx, \quad I_3 = \int_0^{\frac{\pi}{6}} \tan^5 x\,dx.$$

5. Prove that $\dfrac{1}{1+x^2} \geq 1-x^2$. By integration, deduce that if $c > 0$,

$$\tan^{-1} c > c - \frac{c^3}{3}.$$

6. Prove that $\displaystyle\int_0^1 \frac{dx}{\sqrt{(1+x^5)}} \geq \int_0^1 \frac{dx}{\sqrt{(1+x)}}$. Hence, prove that

$$2\sqrt{2}-2 \leq \int_0^1 \frac{dx}{\sqrt{(1+x^5)}} \leq 1.$$

Try to estimate the integral more precisely.

7. Prove that, if $b > a$,

$$\left|\int_a^b f(x)\,dx\right| \leq \int_a^b |f(x)|\,dx.$$

8. By considering the integral $\displaystyle\int_1^t \frac{dx}{x}$, prove that,

if $t > 1$, $\quad t-1 > \log t > \dfrac{t-1}{t}$.

A useful inequality for integrals

If $h \leq f(x) \leq k$ for all values of x in the range $a \leq x \leq b$ and if $g(x)$ is positive throughout this range, then

$$\{f(x)-h\}g(x) \geq 0 \quad \text{and} \quad \{k-f(x)\}g(x) \geq 0$$

throughout this range. Hence
$$\int_a^b \{f(x)-h\}g(x)\,dx \geq 0,$$
and
$$\int_a^b \{k-f(x)\}g(x)\,dx \geq 0,$$
whence
$$h\int_a^b g(x)\,dx \leq \int_a^b f(x)g(x)\,dx \leq k\int_a^b g(x)\,dx.$$

Example. Consider the integral
$$I = \int_1^2 \frac{1}{x^2\sqrt{(1+x^5)}}\,dx.$$
Since $\frac{1}{\sqrt{(1+x^5)}}$ lies between $\frac{1}{\sqrt{33}}$ and $\frac{1}{\sqrt{2}}$ for all values of x lying in the range $1 \leq x \leq 2$,
$$\frac{1}{\sqrt{33}}\int_1^2 \frac{1}{x^2}\,dx \leq I \leq \frac{1}{\sqrt{2}}\int_1^2 \frac{1}{x^2}\,dx,$$
whence
$$\frac{1}{2\sqrt{33}} \leq I \leq \frac{1}{2\sqrt{2}}.$$

Exercises 1(f) (cont.)

9. If m is the least value and M the greatest value of $f(x)$ in the range $a \leq x \leq b$, and if $g(x)$ is positive throughout the range, show that
$$m\int_a^b g(x)\,dx \leq \int_a^b f(x)g(x)\,dx \leq M\int_a^b g(x)\,dx.$$

State and prove the corresponding theorem in the case when $g(x)$ is negative throughout the range.

10. By considering the maximum and minimum values of $\frac{1}{\sqrt{(1+x^3)}}$ in the range, estimate the value of the integral
$$I = \int_1^3 \frac{dx}{x^3\sqrt{(1+x^3)}},$$
and show that $0\cdot08 < I < 0\cdot32$.

CONVERGENCE OF SERIES AND INTEGRALS 27

By dividing up the range into $(1, \frac{3}{2})$, $(\frac{3}{2}, 2)$, $(2, 3)$ obtain a more refined inequality $0{\cdot}17 < I < 0{\cdot}27$.

Finally, by dividing the range into $(1, \frac{5}{4})$, $(\frac{5}{4}, \frac{3}{2})$, $(\frac{3}{2}, \frac{7}{4})$, $(\frac{7}{4}, 2)$, $(2, 3)$, prove that $0{\cdot}200 < I < 0{\cdot}251$.

11. If $f'(x)$ has a constant sign throughout the range $0 \leq x \leq na$, prove that $\int_0^{na} f(x) \, dx$ lies between

$$a\{f(0) + f(a) + f(2a) + \cdots + f(\overline{n-1}a)\},$$

and

$$a\{f(a) + f(2a) + \cdots + f(na)\}.$$

Hence, show that $\int_0^1 (1-x^2)^{\frac{1}{2}} \, dx$ lies between

$$\frac{1}{n} \sum_{r=0}^{n-1} \left(1 - \frac{r^2}{n^2}\right)^{\frac{1}{2}} \quad \text{and} \quad \frac{1}{n} \sum_{r=1}^{n} \left(1 - \frac{r^2}{n^2}\right)^{\frac{1}{2}}.$$

Hence, taking $n = 10$ and using 5-figure square root tables, find limits between which π must lie, making allowance for the accumulated error arising from the approximations in the square roots table.

Show that for any value of n the maximum value of the error in calculating π from the mean of the two results is approximately $\dfrac{2}{n} + 0{\cdot}000\,04$. The first term in the error can be reduced by increasing n but the second term can be improved only by using square root tables of greater accuracy.

Estimating the value of an integral

Consider $I = \displaystyle\int_0^{\frac{1}{2}} \frac{dx}{1-x^2}$, whose value is $\frac{1}{2} \log 3$. By division we can express the integrand as

$$\frac{1}{1-x^2} = 1 + x^2 + x^4 + x^6 + \frac{x^8}{1-x^2},$$

whence

$$I = \frac{1}{2} + \frac{(\frac{1}{2})^3}{3} + \frac{(\frac{1}{2})^5}{5} + \frac{(\frac{1}{2})^7}{7} + \int_0^{\frac{1}{2}} \frac{x^8 \, dx}{1-x^2}.$$

We can estimate the last integral. It is clear that throughout the range of integration

$$\left|\frac{x^8}{1-x^2}\right| < \frac{(\frac{1}{2})^8}{1-(\frac{1}{2})^2} = \frac{1}{3 . 2^7},$$

whence
$$\left| \int_0^{\frac{1}{2}} \frac{x^8}{1-x^2} dx \right| < \frac{1}{3 \cdot 2^7} \frac{1}{2} < 0.002\ 81$$

Thus I lies between
$$\frac{1}{2} + \frac{(\frac{1}{2})^3}{3} + \frac{(\frac{1}{2})^5}{5} + \frac{(\frac{1}{2})^7}{7}$$
and
$$\frac{1}{2} + \frac{(\frac{1}{2})^3}{3} + \frac{(\frac{1}{2})^5}{5} + \frac{(\frac{1}{2})^7}{7} + 0.002\ 81.$$

Hence log 3 lies between $2(0.548\ 82)$ and $2(0.548\ 83 + 0.002\ 81)$, i.e. between 1·097 and 1·104. In fact log 3 is 1·0986 to four decimal places.

Repeat this process taking $\frac{1}{1-x^2} = 1 + x^2 + \cdots + x^{10} + \frac{x^{12}}{1-x^2}$, and find a slimmer interval in which log 3 must lie.

Exercises 1(f) (cont.)

12. Find the limiting values of the following integrals as R tends to infinity.

$$\int_R^{2R} \frac{x}{x^3+1} dx; \quad \int_{2R}^{5R} \frac{1}{x^2-10} dx; \quad \int_R^{3R} \frac{x^2}{x^4+x^2+1} dx;$$

$$\int_R^{2R} \frac{x}{x^3-4x-7} dx; \quad \int_R^{6R} \frac{e^{-x}}{1+x^2} dx; \quad \int_{4R}^{5R} x^3 e^{-2x} dx;$$

$$\int_0^{\frac{\pi}{2}} \frac{e^{-R\sin\theta}}{R^2+1} d\theta.$$

Convergence of infinite series and infinite integrals

We shall now prove a result ascribed to Maclaurin and Cauchy establishing a connection between the convergence of an infinite series and the convergence of an associated infinite integral.

If $f(x)$ is positive for $x \geq 1$ and steadily decreases as x increases, then the series
$$f(1) + f(2) + f(3) + \cdots$$
and the integral $\int_1^\infty f(x) dx$ both converge or both diverge, i.e. the convergence of the integral implies the convergence of the series,

and conversely. The divergence of the integral implies the divergence of the series, and conversely.

In the range $r-1 \leq x \leq r$,

$$f(r) \leq f(x) \leq f(r-1),$$

since $f(x)$ is steadily decreasing as x increases. Hence, by the theorem proved on p. 21

$$f(r) \leq \int_{r-1}^{r} f(x)\,dx \leq f(r-1).$$

Putting $r = 2, 3, \ldots n$, and adding the inequalities, we have

$$f(2)+f(3)+\cdots+f(n) \leq \int_{1}^{n} f(x)\,dx \leq f(1)+f(2)+\cdots+f(n-1).$$

Thus if S_n denotes the sum of the first n terms of the series

$$f(1)+f(2)+\cdots,$$

and I_n denotes $\int_{1}^{n} f(x)\,dx$,

$$S_n - f(1) \leq I_n \leq S_n - f(n).$$

If the integral $\int_{1}^{\infty} f(x)\,dx$ converges, we deduce from the inequality

$$S_n \leq I_n + f(1)$$

that S_n is bounded above (i.e. $S_n < k$ for all values of n), from which it follows that the infinite series

$$f(1)+f(2)+f(3)+\cdots$$

is convergent.

If we know that the infinite series converges, we deduce from the inequality

$$I_n \leq S_n - f(n) \leq S_n,$$

that I_n is bounded, which implies the convergence of the integral.

If the integral is divergent, we deduce from the inequality $S_n \geq I_n + f(n)$ that S_n tends to infinity as n tends to infinity and hence that the series diverges.

If the series is divergent, we deduce from the inequality

$$I_n \geq S_n - f(1)$$

that I_n tends to infinity as n tends to infinity and hence that the integral is divergent.

Convergence of the series $\sum_{n=1}^{\infty} \dfrac{1}{n^s}$

It is obvious that the series diverges if $s \leq 0$, since the nth term does not tend to 0 as $n \to \infty$.

If $s > 0$, x^{-s} is positive and steadily decreases as x increases for $x \geq 1$. It follows that the series converges or diverges with the integral $\int_1^{\infty} x^{-s}\,\mathrm{d}x$. Now the integral converges if $s > 1$ and diverges if $s \leq 1$. Hence the series $\sum n^{-s}$ converges if $s > 1$ and diverges if $s \leq 1$.

Euler's constant

Using the notation of the last paragraph, we shall now prove that $(S_n - I_n)$ must tend to a limit as n tends to infinity. That this must happen when the series and the integral are convergent is obvious but the point is that it is also true when the series and the integral are divergent.

Consider the sequence whose nth term is $S_n - I_n$. First, we may prove that the sequence is steadily decreasing; for

$$(S_n - I_n) - (S_{n+1} - I_{n+1}) = (I_{n+1} - I_n) - (S_{n+1} - S_n)$$
$$= \int_n^{n+1} f(x)\,\mathrm{d}x - f(n+1) \geq 0.$$

Further, the sequence is bounded below. For $S_n - I_n \geq f(n) \geq 0$.

From these two facts, it follows that the terms of the sequence must tend to a limit as n tends to infinity; i.e. $S_n - I_n$ tends to a limit as n tends to infinity.

Example. The series $1 + \tfrac{1}{2} + \tfrac{1}{3} + \cdots$ and the integral $\int_1^{\infty} \dfrac{\mathrm{d}x}{x}$ diverge. But

$$S_n - I_n = 1 + \frac{1}{2} + \frac{1}{3} + \cdots + \frac{1}{n} - \log n$$

tends to a limit, known as Euler's constant γ, as n tends to infinity.

Exercises 1(g)

1. Prove that the series $\sum_{n=2}^{\infty} \dfrac{1}{n(\log n)^s}$ converges if $s > 1$ and diverges if $s \leq 1$.

CONVERGENCE OF SERIES AND INTEGRALS

2. Find an approximate value for Euler's constant γ, correct to 4 decimal places.

3. If S_n denotes $\sum_{r=1}^{n} \frac{1}{r}$ and k is an integer, prove that $\lim_{n \to \infty} \frac{S_{n+k}}{S_n} = 1$ and that $\lim_{n \to \infty} \frac{S_{(n^k)}}{S_n} = k$.

4. Prove that the series $\sum_{n=1}^{\infty} \left[\frac{1}{n} - \log\left(1 + \frac{1}{n}\right) \right]$ is convergent.

Example. Prove that the series

$$1 - \tfrac{1}{2} + \tfrac{1}{3} - \tfrac{1}{4} + \cdots$$

converges and find its sum.

If S_n denotes the sum of the first n terms of the series,

$$\begin{aligned}
S_{2n} &= \left(1 + \frac{1}{3} + \frac{1}{5} + \cdots + \frac{1}{2n-1}\right) - \left(\frac{1}{2} + \frac{1}{4} + \cdots + \frac{1}{2n}\right) \\
&= \left(1 + \frac{1}{2} + \frac{1}{3} + \cdots + \frac{1}{2n}\right) - 2\left(\frac{1}{2} + \frac{1}{4} + \cdots + \frac{1}{2n}\right) \\
&= \left(1 + \frac{1}{2} + \frac{1}{3} + \cdots + \frac{1}{2n} - \log(2n)\right) - \\
&\quad - \left(1 + \frac{1}{2} + \frac{1}{3} + \cdots + \frac{1}{n} - \log n\right) + \log 2.
\end{aligned}$$

Each of the two brackets on the right-hand side tends to γ as n tends to infinity. It follows that S_{2n} tends to the value $\log 2$ as n tends to infinity. Further,

$$S_{2n+1} = S_{2n} + \frac{1}{2n+1},$$

and hence S_{2n+1} also tends to $\log 2$ as n tends to infinity. Thus S_n tends to $\log 2$ as n tends to infinity. Hence the series converges and has sum $\log 2$.

If the terms in the series are rearranged so that two positive terms are followed by one negative term, giving the series

$$1 + \tfrac{1}{3} - \tfrac{1}{2} + \tfrac{1}{5} + \tfrac{1}{7} - \tfrac{1}{4} + \cdots,$$

S_{3n} may be written in the form

$$S_{3n} = \left\{1+\frac{1}{2}+\frac{1}{3}+\cdots+\frac{1}{4n}-\log(4n)\right\}-$$
$$-\frac{1}{2}\left\{1+\frac{1}{2}+\frac{1}{3}+\cdots+\frac{1}{2n}-\log(2n)\right\}-$$
$$-\frac{1}{2}\left\{1+\frac{1}{2}+\frac{1}{3}+\cdots+\frac{1}{n}-\log n\right\}+\tfrac{1}{2}\log 8.$$

∴ $S_{3n} \to \tfrac{1}{2}\log 8$ as $n \to \infty$. S_{3n+1} and S_{3n+2} tend to the same limit and thus the series is convergent and has sum $\tfrac{1}{2}\log 8$.

This example demonstrates that if a convergent series is deranged, the resulting series does not necessarily converge to the same sum as the original series. We shall now investigate the general problem of derangement.

Sum of a deranged series

Let Σu_n be a convergent series and Σv_n a series obtained by deranging the terms of Σu_n.

Consider first the case when u_n is positive for all values of n; denote by S_n the sum of n terms of the series Σu_n and by S'_n the sum of n terms of the series Σv_n. Then for any value of n, there exists a value N such that S_N contains all the terms of S'_n, whence $S'_n < S_N$. But since Σu_n converges, we can find a positive K such that $S_N < K$ for all values of N, and hence $S'_n < K$ for all values of n. It follows that the series Σv_n converges. Further, if $S = \sum_{n=1}^{\infty} u_n$ and $S' = \sum_{n=1}^{\infty} v_n$, we have, on taking the limit as $n \to \infty$ in $S'_n < S_N$ that $S' \leq S$. But Σu_n can be regarded as a derangement of Σv_n whence $S \leq S'$. It follows that $S' = S$. Thus if a convergent series of positive terms is deranged its sum remains unchanged.

Next suppose that u_n may have any sign but the series Σu_n converges absolutely.

Let
$$a_n = \tfrac{1}{2}(|u_n|+u_n), \qquad b_n = \tfrac{1}{2}(|u_n|-u_n),$$
$$c_n = \tfrac{1}{2}(|v_n|+v_n), \qquad d_n = \tfrac{1}{2}(|v_n|-v_n).$$

Then Σa_n, Σb_n, Σc_n, Σd_n are series of positive terms, Σc_n being a

CONVERGENCE OF SERIES AND INTEGRALS 33

derangement of $\Sigma\, a_n$, and $\Sigma\, d_n$ a derangement of $\Sigma\, b_n$. Now $\Sigma\, a_n$ converges and hence $\Sigma\, c_n$ converges; further $\sum_{n=1}^{\infty} a_n = \sum_{n=1}^{\infty} c_n$. Similarly $\sum_{n=1}^{\infty} b_n = \sum_{n=1}^{\infty} d_n$. Hence

$$\sum_{n=1}^{\infty}(a_n - b_n) = \sum_{n=1}^{\infty}(c_n - d_n),$$

i.e.

$$\sum_{n=1}^{\infty} u_n = \sum_{n=1}^{\infty} v_n.$$

Thus, if we derange the terms of an absolutely convergent series, the resulting series also converges to the same sum as the original series, but if we derange the terms of a conditionally convergent series, i.e. one which is not absolutely convergent, then the resulting series may converge to a different sum from the original series, or may diverge.

Exercises 1(h)

1. Show that the series obtained by rearranging the terms of the series

$$1 - \tfrac{1}{2} + \tfrac{1}{3} - \tfrac{1}{4} + \cdots$$

so that three negative terms follow one positive term is convergent and has sum $\tfrac{1}{2} \log(\tfrac{4}{3})$.

2. Prove that the sum of n terms of the series

$$1 + \tfrac{1}{2} - \tfrac{1}{3} + \tfrac{1}{4} + \tfrac{1}{5} - \tfrac{1}{6} + \tfrac{1}{7} + \tfrac{1}{8} - \tfrac{1}{9} + \cdots$$

tends to infinity as n tends to infinity, but that the infinite series

$$1 + \tfrac{1}{3} + \tfrac{1}{5} - \tfrac{1}{2} + \tfrac{1}{7} + \tfrac{1}{9} + \tfrac{1}{11} - \tfrac{1}{4} + \tfrac{1}{13} + \cdots$$

converges.

3. State whether the following series converge and if so, find their sums:

(i) $1 - \dfrac{1}{2^2} + \dfrac{1}{3^2} - \dfrac{1}{4^2} + \dfrac{1}{5^2} - \cdots$

(ii) $1 + \dfrac{1}{3^2} - \dfrac{1}{2^2} + \dfrac{1}{5^2} + \dfrac{1}{7^2} - \dfrac{1}{4^2} + \cdots$

(iii) $-\dfrac{1}{2^2} + 1 + \dfrac{1}{3^2} + \dfrac{1}{5^5} - \dfrac{1}{4^2} + \dfrac{1}{7^2} + \dfrac{1}{9^2} + \dfrac{1}{11^2} - \cdots$

4. Explain how to derange the series $\sum_{n=1}^{\infty} (-1)^{n-1}/n$ to give a convergent series of sum 4.

5. If the series $1-\tfrac{1}{2}+\tfrac{1}{3}-\tfrac{1}{4}+\cdots$ is deranged so that p positive terms are followed by q negative terms throughout, prove that the resulting series is convergent and has sum $\log 2 + \tfrac{1}{2}\log(p/q)$.

6. Prove that the series
$$\frac{1}{\sqrt{1}}-\frac{1}{\sqrt{2}}+\frac{1}{\sqrt{3}}-\frac{1}{\sqrt{4}}+\cdots$$
converges, but that the deranged series obtained by taking two positive terms followed by one negative term, i.e. the series
$$\frac{1}{\sqrt{1}}+\frac{1}{\sqrt{3}}-\frac{1}{\sqrt{2}}+\frac{1}{\sqrt{5}}+\frac{1}{\sqrt{7}}-\frac{1}{\sqrt{4}}+\cdots$$
diverges. (*Hint.* Prove that $\dfrac{1}{\sqrt{(4p-3)}}+\dfrac{1}{\sqrt{(4p-1)}}-\dfrac{1}{\sqrt{(2p)}}$ is greater than $\dfrac{\sqrt{2}-1}{\sqrt{(2p)}}$.

7. Prove that if the terms of the series $\sum_{n=1}^{\infty}(-1)^{n-1}n^{-\tfrac{3}{2}}$ are deranged the resulting series converges to the same sum as the original series.

Derange the terms of the series $\sum(-1)^{n-1}n^{-\tfrac{1}{3}}$ in such a way that the resulting series diverges.

Miscellaneous problems on Chapter 1

1. Sketch the curve $y = \dfrac{(x-1)^2}{x^4}$ and prove that the area contained between the curve and the portion of the x-axis for which $x \geq 1$ is $\tfrac{1}{3}$.

2. Discuss the convergence for all values of z, real and complex, of the series whose nth terms are
$$(n+1)^n z^n; \qquad n^2(z-1)^n; \qquad \frac{n^3}{(n+1)!}z^n.$$

3. Prove that if $t > 0$, and $S_n = 1^t + 2^t + \cdots + n^t$, then
$$S_n > \int_0^n x^t\,dx > S_n - n^t.$$
Hence prove that the limiting value of $\dfrac{S_n}{1+t}$ as n tends to infinity is $\dfrac{1}{t+1}$. Illustrate in the case $t = 3$ by evaluating the limit in an elementary way.

4. Investigate the convergence or divergence of the following integrals:
$$\int_{-1}^{1} x^{-(2s+1)}\,dx; \qquad \int_0^3 (x-2)^{-\tfrac{3}{5}}\,dx; \qquad \int_0^4 (x-1)^{-1}(x-2)^{-\tfrac{2}{3}}\,dx.$$

5. Prove that the integral $\int_0^\pi \frac{\cos x}{\sqrt{(\sin x)}} dx$ is convergent and evaluate it.

6. Prove that the infinite series
$$1 + \tfrac{1}{3} + \tfrac{1}{5} + \tfrac{1}{7} - \tfrac{1}{2} + \tfrac{1}{9} + \tfrac{1}{11} + \tfrac{1}{13} + \tfrac{1}{15} - \tfrac{1}{4} + \cdots$$
is convergent with sum log 4.

7. Evaluate the following integrals:
$$\int_0^\infty \frac{t^2 \, dt}{(1+t^2)^4}; \quad \int_0^\infty \frac{x \, dx}{x^3+1}; \quad \int_2^\infty \frac{dx}{(1+x)^2(1+x^2)}.$$

8. Prove that $\int_1^\infty \frac{dx}{x\{1+(\log x)^3\}}$ converges. Under what condition does the integral $\int_3^\infty \frac{dx}{x(\log x)(\log \log x)^s}$ converge?

9. Discuss the convergence of the following series:

(i) $\sum_{n=1}^\infty \frac{n+2\sqrt{n}}{(n^2+3)(2n-1)}$;

(ii) $1 - \dfrac{1}{2^k} + \dfrac{1}{3^k} - \cdots$ for real values of k.

10. Find the real values of x for which the series
$$x + \frac{x^3}{3} - \frac{x^2}{2} + \frac{x^5}{5} + \frac{x^7}{7} - \frac{x^4}{4} + \cdots$$
converges and find its sum when it converges.

11. Prove that if $a > b > 0$, then
$$\int_{-\infty}^\infty \frac{d\theta}{a \cosh \theta + b \sinh \theta} = \frac{\pi}{\sqrt{(a^2-b^2)}}.$$

12. For what values of k does the integral $\int_0^\pi \frac{dx}{1+k \sin x}$ converge? Evaluate the integral when it converges.

13. Discuss the convergence of the following series for real values of x (u_n denotes the nth term of the series).

(i) $u_n = \dfrac{n^3 x^n}{n^2+1}$; (ii) $u_n = \dfrac{1}{1+nx^n}$;

(iii) $u_n = \dfrac{n(n+1)x^{n+2}}{(n+2)(n+3)(n+4)}$; (iv) $u_n = \dfrac{n}{(n+1)^2}(-x)^{n+1}$.

14. Prove that $\sum_{n=1}^{\infty} \dfrac{1}{n(4n^2-1)} = 2\log 2 - 1$.

15. By dividing up the range of integration in $\int_{4}^{8} \dfrac{dx}{x}$ into intervals of length 1 and estimating the integrals so formed, show that $\log 2$ lies between 0·63 and 0·76. Obtain a more accurate estimate of $\log 2$ by dividing the range of integration into intervals of length $\tfrac{1}{2}$.

16. If Σu_n is a series of positive terms such that, for all values of n from a certain stage,
$$\frac{u_n}{u_{n+1}} = 1 + \frac{a}{n} + \frac{b}{n^2},$$
where a is independent of n and b is a bounded function of n, then the series will converge or diverge according as $a > 1$ or $a \leq 1$.

If λ and k are positive, prove that the series
$$1 + \frac{\lambda^2}{1!k} + \frac{\lambda^2(\lambda+1)^2}{2!k(k+1)} + \frac{\lambda^2(\lambda+1)^2(\lambda+2)^2}{3!k(k+1)(k+2)} + \cdots$$
converges if $k > 2\lambda$ and diverges if $k \leq 2\lambda$.

17. For each of the series below, state whether the final result is true, and whether the argument given is correct or not.

 (i) In the series $\Sigma (\tfrac{1}{4})^n$, $u_n \to 0$, hence the series converges.

 (ii) in the series $\Sigma \dfrac{(-1)^n}{n+3}$, the terms alternate in sign; hence the series converges.

 (iii) in the series $\Sigma \left(1 + \dfrac{1}{n^2}\right)$, u_n does not tend to 0; hence the series diverges.

 (iv) If Σa_n is a convergent series of real terms and $b_n < a_n$, then Σb_n converges.

18. Prove that the integral $\int_{0}^{\infty} \dfrac{x^2\,dx}{(1+x^2)^5}$ converges, and that its value is $\dfrac{5\pi}{256}$.

19. Find the sum of the series $\sum_{n=1}^{\infty} \dfrac{n^3 z^{n-1}}{(n-1)!}$ for real values of z, and state whether the series converges for complex values of z.

20. Find all the values of the real variable x for which the following series are absolutely convergent:
$$\Sigma \frac{x^{n+1}}{n+1}; \qquad \Sigma \frac{x^n(1-x)^{n+1}}{n+1}.$$

CONVERGENCE OF SERIES AND INTEGRALS 37

21. Prove that if $0 < k < 1$,
$$\int_0^\infty \frac{dx}{k^2 + \sinh^2 x} = \frac{1}{k} \frac{\cos^{-1} k}{\sqrt{(1-k^2)}},$$
and find the value of the integral if $k > 1$.

22. Prove that the series
$$\left(\frac{t}{1+t^2}\right) + \sum_{n=2}^\infty \frac{(2n-2)!}{n[(n-1)!]^2} \left(\frac{t}{1+t^2}\right)^{2n-1}$$
converges to t if $|t| < 1$, and to $\frac{1}{t}$ if $|t| > 1$. Discuss the convergence of the series in the case $|t| = 1$.

23. Discuss the convergence or divergence of the following series Σu_n, where
$$u_n = \frac{(-1)^{n-1}}{\sqrt{n + (-1)^{n-1}}} \ ; \qquad u_n = \frac{\sin^2 n\theta}{n}.$$

24. By dividing the range of integration in $\int_0^2 \frac{dx}{1+x^2}$ into intervals of width $\frac{1}{2}$, prove that $\tan^{-1} 2$ lies between 0·90 and 1·31 and find a closer approximation.

25. Discuss the convergence or divergence of the following series:
$$\Sigma (-1)^n \sin\left(\frac{\pi}{2n}\right); \qquad \Sigma \frac{1}{n(n+2)} \cos\left(\frac{\pi}{4n}\right);$$
$$\Sigma (-1)^n n^{-1/n}; \qquad \Sigma \frac{n \log n}{n^2 + 2}.$$

26. If the sequence (a_n) of non-negative real numbers is steadily decreasing and $\lim_{n\to\infty} a_n = 0$, prove that if r_m denotes the remainder after m terms of the series $\sum_1^\infty (-1)^n a_n$, then $|r_m| < a_{m+1}$.

By taking sufficiently many terms to ensure that the error in approximating to $\sum_1^\infty (-1)^n/n^3$ is less than 0·01, show that
$$-0.995 < \sum_1^\infty (-1)^n/n^3 < -0.974.$$

By considering the sign of the error term show that the upper estimate can be improved to -0.984.

27. If (a_n) is a steadily decreasing sequence of positive terms and Σa_n is convergent, prove by considering $r_m = \sum_{t=m+1}^{2m} a_t$ that na_n tends to 0 as n tends to infinity.

28. If $u_n > 0$ and the series Σu_n converges, prove that the series Σu_n^4 and $\Sigma \dfrac{u_n}{1+u_n}$ also converge.

29. If P_n denotes $(1+u_1)(1+u_2) \cdots (1+u_n)$, and P_n tends to a unique limit other than zero as n tends to infinity, we call the infinite product $(1+u_1)(1+u_2)\ldots$, denoted by $\prod_{n=1}^{\infty}(1+u_n)$, convergent. Prove that the infinite product is convergent if the series $\sum_{n=1}^{\infty} u_n$ is absolutely convergent, in which case we say that the infinite product is absolutely convergent.

Prove that the infinite product $\prod_{r=1}^{\infty}\left(1-\dfrac{x^2}{r^2\pi^2}\right)$ is absolutely convergent; it may be proved that its value is $\dfrac{\sin x}{x}$. Hence prove that $\sum_{n=1}^{\infty}\dfrac{1}{n^2} = \dfrac{\pi^2}{6}$ and that $\prod_{p \text{ prime}}\left(1-\dfrac{1}{p^2}\right)^{-1}$ has the same value.

30. Prove that if $x \geq 3$, $\dfrac{\log x}{x}$ steadily decreases as x increases, and deduce that the sequence $(n^{1/n})$ is monotonic decreasing.

If Σa_n is a convergent series of positive terms, prove that the series $\Sigma n^{1/n} a_n$ also converges and investigate the behaviour of the series $\Sigma n^{1/n-1}$.

31. Prove that $f(n) = \sum_{r=1}^{n}\dfrac{n}{n^2+r^2}$ lies between $\int_0^{n-1}\dfrac{n}{n^2+x^2}dx$ and $\int_1^{n}\dfrac{n}{n^2+x^2}dx$. Hence show that $f(n)$ tends to $\tfrac{1}{4}\pi$ as n tends to infinity.

32. If $f(x) = \dfrac{8+x-\tfrac{1}{18}x^2}{\sqrt{(4+x)}}$, prove that $4 \leq f(x) \leq 4{\cdot}0005$ in the interval $[0, \tfrac{1}{2}]$. Hence prove that

$$1{\cdot}0332 \leq \int_0^{\frac{1}{2}} \dfrac{8+x-\tfrac{1}{18}x^2}{\sqrt{(16-x^2)}}dx \leq 1{\cdot}0338,$$

and deduce that $0{\cdot}1252 < \sin^{-1}(\tfrac{1}{8}) < 0{\cdot}1254$.

33. Prove that, if r_m is the error in approximating to $S = \sum_{n=1}^{\infty}\dfrac{(-1)^{n+1}}{n^4}$ by summing the first m terms of the series, then $|r_m| < \dfrac{1}{(m+1)^4}$.

Choose m large enough to make $|r_m| < 10^{-3}$ and hence show that $0{\cdot}9467 < S < 0{\cdot}9484$.

34. State which of the following conditions are sufficient to imply that

the series $\sum u_n$ is convergent. If the condition is not sufficient give either a reason or an example to justify your answer.
 (i) $\sum u_n^2$ is convergent.
 (ii) $u_n > 0$ and $\dfrac{u_{n+1}}{u_n} < 1$ for all values of n.
 (iii) $u_n > 0$ and $\sum \log u_n$ is convergent.
 (iv) $u_n > 0$ and $u_n^{\frac{1}{n}} < k < 1$ for all values of n.
 (v) $u_n < 0$ and $\sum e^{u_n}$ is convergent.
 (vi) $nu_n \to 0$ as $n \to \infty$.
 (vii) u_n is alternately positive and negative and $u_n \to 0$ as $n \to \infty$.
 (viii) $\sum n^2 u_n$ is convergent.

35. Prove that if $\sum u_n$ is convergent, then $\sum (u_n + u_{n+1})$ is also convergent. Give an example to show that the converse is not true.

2
Power Series

Differentiation of finite series

If $y = y_1 + y_2 + \cdots + y_n$, where $y_1, y_2, \ldots$ are differentiable functions of x and n is finite, then $\dfrac{dy}{dx}$ may be obtained by term-by-term differentiation, i.e.

For
$$\frac{dy}{dx} = \frac{dy_1}{dx} + \frac{dy_2}{dx} + \cdots + \frac{dy_n}{dx}.$$

$$\frac{\delta y}{\delta x} = \frac{\delta y_1}{\delta x} + \frac{\delta y_2}{\delta x} + \cdots + \frac{\delta y_n}{\delta x}.$$

Now, given any small positive ε, we can choose an appropriate small ε_1 so that if $|\delta x| < \varepsilon_1$ each term on the right-hand side $\dfrac{\delta y_r}{\delta x}$ differs from $\dfrac{dy_r}{dx}$ by an amount whose modulus is less than ε. Thus,

differs from
$$\frac{\delta y_1}{\delta x} + \frac{\delta y_2}{\delta x} + \cdots + \frac{\delta y_n}{\delta x}$$
$$\frac{dy_1}{dx} + \frac{dy_2}{dx} + \cdots + \frac{dy_n}{dx}$$

by an amount whose modulus† is less than $n\varepsilon$, and provided n is finite this implies that

† For all values of v_r, real or complex, it is easily proved that
$$|v_1 + v_2 + \cdots + v_n| \leq |v_1| + |v_2| + \cdots + |v_n|.$$

$$\lim_{\delta x \to 0} \left(\frac{\delta y_1}{\delta x} + \frac{\delta y_2}{\delta x} + \cdots + \frac{\delta y_n}{\delta x} \right) = \frac{dy_1}{dx} + \frac{dy_2}{dx} + \cdots + \frac{dy_n}{dx},$$

which proves the theorem.

Differentiation of infinite series

If $y = y_1 + y_2 + \ldots$, a convergent infinite series of differentiable functions of x, it is not always true that $\dfrac{dy}{dx}$ can be obtained by term-by-term differentiation. For example, consider the infinite series
$$x - \frac{x^2}{2} + \frac{x^3}{3} - \cdots,$$
which converges if $-1 < x \leq 1$. Term-by-term differentiation of the infinite series gives
$$1 - x + x^2 - \cdots,$$
which converges if $-1 < x < 1$, but diverges if $x = 1$. Thus, the process of term-by-term differentiation of a convergent series may lead to a divergent series, and even if it leads to a convergent series, the sum of this latter series is not necessarily equal to dy/dx. We shall not deal in this volume with conditions of validity for term-by-term differentiation of a general series of the form $\sum\limits_{r=1}^{\infty} f_r(x)$, but we shall investigate the conditions for an important class of series, namely power series.

Differentiation of power series

We shall now prove two important theorems about power series. Consider the power series
$$f(x) = a_0 + a_1 x + a_2 x^2 + \cdots,$$
and let us assume that the series converges absolutely for $|x| < r$.

We shall first prove that, if $|x| < r$, $f'(x)$ can be found by term-by-term differentiation of the series, i.e.
$$f'(x) = a_1 + 2a_2 x + 3a_3 x^2 + \cdots.$$

Secondly, we shall prove that if (a, b) is any interval within the range $|x| < r$, $\int_a^b f(x)\, dx$ can be found by term-by-term integration of the series, i.e.
$$\int_a^b f(x)\, dx = [a_0 x]_a^b + \left[\frac{a_1 x^2}{2}\right]_a^b + \left[\frac{a_2 x^3}{3}\right]_a^b + \cdots.$$

These results are what we should expect but the reader will be aware of the dangers that lie in the manipulation of infinite series and will recognise the necessity for formal proofs of the results.

We shall begin by proving that the series

$$a_1 + 2a_2 x + 3a_3 x^2 + \cdots$$

will converge absolutely if $|x| < r$.

For, if $|x| < t < r$, (where t is any positive number less than r), the absolute convergence of $\Sigma a_n t^n$ implies that N exists such that, if $n > N$, $|a_n t^n| < 1$. Thus, disregarding the first N terms of the series which do not affect the convergence of the series, $|a_n| < \dfrac{1}{t^n}$ and hence $|na_n x^n| < n \left|\dfrac{x}{t}\right|^n$. Now $\Sigma n \left|\dfrac{x}{t}\right|^n$ converges absolutely by D'Alembert's test and hence $\Sigma n a_n x^n$ converges absolutely.

Let x and x_1 be values lying within the range $(-r, r)$ so that $|x| < t < r$, $|x_1| < t < r$. Then

$$f(x_1) - f(x) = a_1(x_1 - x) + a_2(x_1^2 - x^2) + \cdots,$$

and

$$\frac{f(x_1) - f(x)}{x_1 - x} = a_1 + a_2(x_1 + x) + a_3(x_1^2 + x_1 x + x^2) + \cdots +$$

$$+ a_n(x_1^{n-1} + x_1^{n-2} x + \cdots + x^{n-1}) + R,$$

where

$$|R| < (n+1)|a_{n+1}| t^n + (n+2)|a_{n+2}| t^{n+1} + \cdots.$$

Taking the limit as x_1 tends to x, we have

$$f'(x) = a_1 + 2a_2 x + 3a_3 x^2 + \cdots + na_n x^{n-1} + \lim_{x_1 \to x} R.$$

Hence, if we denote by S_n the sum of the first n terms of the series

$$a_1 + 2a_2 x + 3a_3 x^2 + \cdots,$$

$$|f'(x) - S_n| < (n+1)|a_{n+1}| t^n + (n+2)|a_{n+2}| t^{n+1} + \cdots.$$

The expression on the right-hand side of this inequality is the remainder after n terms of the series $\Sigma |na_n t^{n-1}|$ which we have shown to be absolutely convergent; hence the expression on the right-hand side will tend to zero as n tends to infinity. It follows that $f'(x) - S_n$ tends to 0, or that S_n tends to $f'(x)$.

We have thus justified the process of term-by-term differentiation of a power series.

Integration of power series

Let (a, b) be an interval lying within the range $(-r, r)$ so that $|a| < t < r$, $|b| < t < r$. Then we may write the power series in the form
$$f(x) = a_0 + a_1 x + a_2 x^2 \cdots + a_{n-1} x^{n-1} + R_n(x),$$
whence
$$\int_a^b f(x)\,dx = [a_0 x]_a^b + \left[\frac{a_1 x^2}{2}\right]_a^b + \cdots + \left[a_{n-1}\frac{x^n}{n}\right]_a^b + \int_a^b R_n(x)\,dx.$$

If we can now prove that
$$\lim_{n \to \infty} \int_a^b R_n(x)\,dx = 0,$$
it will follow that the series
$$[a_0 x]_a^b + \left[a_1 \frac{x^2}{2}\right]_a^b + \cdots$$
is convergent, and that the sum of the series is $\int_a^b f(x)\,dx$. Clearly
$$|R_n(x)| < |a_n| t^n + |a_{n+1}| t^{n+1} + \cdots,$$
whence
$$\left|\int_a^b R_n(x)\,dx\right| < \{|a_n| t^n + |a_{n+1}| t^{n+1} + \cdots\}\{|b-a|\}.$$

Now the first bracket represents the remainder after n terms of an absolutely convergent series and therefore tends to 0 as n tends to ∞, and the second bracket is finite. It follows that
$$\lim_{n \to \infty} \int_a^b R_n(x)\,dx = 0.$$

We have thus justified the process of term-by-term integration of a power series.

nth derivative of a function

If $y = f(x)$ is a differentiable function of x, it may happen that dy/dx, $d^2y/dx^2,...$ are also differentiable functions of x, i.e. we can differentiate $f(x)$ any number of times. For example, since e^x, which is differentiable for all values of x, is unchanged by differentiation, it is clearly differentiable any number of times, the result being e^x in all cases.

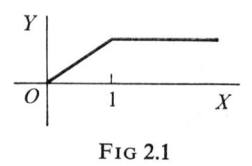

FIG 2.1

But it is not true that all functions can be differentiated any number of times. Consider the function defined by the equations

$$y = 1+x^2 \quad \text{if} \quad x \leq 1;$$
$$y = 2x \quad \text{if} \quad x > 1.$$

This function is differentiable for all values of x, including the critical value $x = 1$; the value of the differential coefficient being given by the equations

$$\frac{dy}{dx} = 2x \quad \text{if} \quad x \leq 1,$$

$$\frac{dy}{dx} = 2 \quad \text{if} \quad x > 1.$$

The graph of dy/dx is shown in Fig. 2.1. Although dy/dx is continuous at $x = 1$, it is not differentiable there.

For some functions $f(x)$, we can find a formula for the nth differential coefficient $\dfrac{d^n}{dx^n} f(x)$.

Examples.

$$\frac{d^n}{dx^n}(e^x) = e^x; \qquad \frac{d^n}{dx^n}(e^{kx}) = k^n e^{kx}.$$

$$\frac{d^n}{dx^n}(\sin x) = \sin x \quad \text{or} \quad \cos x \quad \text{or} \quad -\sin x \quad \text{or} \quad -\cos x,$$

according as n is of the form $4k, 4k+1, 4k+2, 4k+3$. These results may be combined into a single form as follows:

$$\frac{d}{dx}(\sin x) = \cos x = \sin(x+\tfrac{1}{2}\pi),$$

it follows that:

$$\frac{d^2}{dx^2}(\sin x) = \sin[(x+\tfrac{1}{2}\pi)+\tfrac{1}{2}\pi]\frac{d(x+\tfrac{1}{2}\pi)}{dx} = \sin\left(x+\frac{2\pi}{2}\right).$$

In this way we see that $\dfrac{d^n}{dx^n}(\sin x) = \sin\left(x+\dfrac{n\pi}{2}\right)$, a result which may be proved formally by induction.

Exercises 2(a)

1. Consider the function defined by the equations

$$y = x^3 \quad \text{if } x \leq 2;$$
$$y = 3x^2 - 4 \quad \text{if } x > 2.$$

Show that dy/dx exists for all values of x but that d^2y/dx^2 does not exist when $x = 2$.

2. Consider the function defined by the equations

$$y = 2\sin x \quad \text{if } x < 0;$$
$$y = \sin 2x \quad \text{if } x \geq 0.$$

Show that dy/dx and d^2y/dx^2 exist for all values of x, but that d^3y/dx^3 does not exist at $x = 0$.

3. Find the nth differential coefficients of the following functions:

$$\sin ax; \quad \cos x; \quad \cos bx; \quad \sin(ax+b); \quad \cos(ax+b).$$

4. Find the nth differential coefficients of

$$\sin x \sin 3x; \quad \cos 2x \cos 3x; \quad \sin 4x \cos x.$$

5. Find the nth differential coefficients of the following functions:

$$\log x; \quad \log(ax+b); \quad \frac{1}{(x+1)(x+2)}; \quad \frac{2x}{(x+3)(x+5)}.$$

6. Show that, for all values of p,

$$\frac{d^n}{dx^n}(x^p) = p(p-1)(p-2)\cdots(p-n+1)x^{p-n}.$$

Find the nth differential coefficient of $(ax+b)^p$.

Example.

$$\frac{d}{dx}(e^x \sin x) = e^x(\sin x + \cos x) = e^x\sqrt{2}\sin\left(x+\frac{\pi}{4}\right).$$

It follows that

$$\frac{d^2}{dx^2}(e^x \sin x) = \sqrt{2}\,e^x\left\{\sin\left(x+\frac{\pi}{4}\right) + \cos\left(x+\frac{\pi}{4}\right)\right\}$$

$$= (\sqrt{2})^2\, e^x \sin\left(x+\frac{2\pi}{4}\right).$$

In general,

$$\frac{d^n}{dx^n}(e^x \sin x) = (\sqrt{2})^n e^x \sin\left(x+\frac{n\pi}{4}\right),$$

a result which may be proved formally by induction.

We write the nth differential coefficient of $f(x)$ as $f^{(n)}(x)$, which should not be confused with $f^n(x)$, which means the nth power of $f(x)$. $f^{(n)}(a)$ means the value of $f^{(n)}(x)$ when $x = a$. In addition to this notation, we shall continue to use the dash notation for small values of n: $f^{(1)}(x) = f'(x)$, $f^{(2)}(x) = f''(x)$.

Exercises 2(b)

1. (i) $f(x) = \cos 3x$. Find $f^{(n)}(x)$ and $f^{(n)}(0)$.

    (ii) $f(x) = \sin 4x$. Find $f^{(n)}(x)$ and $f^{(n)}\left(\frac{\pi}{4}\right)$.

    (iii) $f(x) = e^x \sin(\sqrt{3}x)$. Find $f^{(n)}(0)$.

2. If $y = e^{ax} \sin bx$, show that $\dfrac{d^n y}{dx^n}$ can be written in the form

$$r^n e^{ax} \sin(bx+n\theta),$$

where $r = \sqrt{(a^2+b^2)}$ and θ is the angle given by $r \sin\theta = b$, $r \cos\theta = a$.

3. If $y = e^{ax} \cos bx$, find an expression for $\dfrac{d^n y}{dx^n}$.

4. Prove that

$$\frac{d^n}{dx^n}\left(\frac{1}{x^2+2x+2}\right) = (-1)^n n!\, \sin(n+1)\theta\,\sin^{n+1}\theta,$$

where $\cot\theta = x+1$.

5. Prove that

$$\frac{d^n}{dx^n}\{e^{ax}\cos(bx+c)\} = r^n e^{ax}\cos(bx+c+n\phi),$$

where $r = \sqrt{(a^2+b^2)}$, $\phi = \tan^{-1}\left(\frac{b}{a}\right)$.

6. Prove that $\frac{d^n}{dx^n}(\tan^{-1} x) = (-1)^{n-1}(n-1)!\ \sin^n \phi \sin n\phi$, where $x = \cot \phi$.

Exercises 2(c)

1. If u and v are differentiable functions of x, we know that

$$\frac{d}{dx}(uv) = u\frac{dv}{dx} + v\frac{du}{dx}.$$

Differentiate this equation and find expressions for $\frac{d^2}{dx^2}(uv)$ and $\frac{d^3}{dx^3}(uv)$. Compare the manner in which the coefficients in these expressions are formed with the way in which the coefficients in $(1+t)(1+t)$ and

$$(1+t)(1+t)(1+t)$$

are formed. Hence write down $\frac{d^4}{dx^4}(uv)$.

Leibniz's theorem

We shall prove by induction the theorem of Leibniz, which gives an expression for the nth differential coefficient of a product uv, where u and v are functions of x, namely,

$$\frac{d^n}{dx^n}(uv) = u\frac{d^n v}{dx^n} + (^nC_1)\frac{du}{dx}\frac{d^{n-1}v}{dx^{n-1}} + \cdots + (^nC_r)\frac{d^r u}{dx^r}\frac{d^{n-r}v}{dx^{n-r}} + \cdots + v\frac{d^n u}{dx^n},$$

where $^nC_r = \dfrac{n!}{r!(n-r)!}$. Let us assume the theorem true when $n = p$. Then

$$\frac{d^p}{dx^p}(uv) = u\frac{d^p v}{dx^p} + (^pC_1)\frac{du}{dx}\frac{d^{p-1}v}{dx^{p-1}} + \cdots + v\frac{d^p u}{dx^p}.$$

On differentiating this equation, we have

$$\frac{d^{p+1}}{dx^{p+1}}(uv) = \left\{u\frac{d^{p+1}v}{dx^{p+1}} + \frac{du}{dx}\frac{d^p v}{dx^p}\right\} + {}^pC_1\left\{\frac{du}{dx}\frac{d^p v}{dx^p} + \frac{d^2 u}{dx^2}\frac{d^{p-1}v}{dx^{p-1}}\right\} + \cdots +$$

$$+ {}^pC_{r-1}\left\{\frac{d^{r-1}u}{dx^{r-1}}\frac{d^{p-r+2}v}{dx^{p-r+2}} + \frac{d^r u}{dx^r}\frac{d^{p-r+1}v}{dx^{p-r+1}}\right\} +$$

$$+ {}^pC_r\left\{\frac{d^r u}{dx^r}\frac{d^{p-r+1}v}{dx^{p-r+1}} + \frac{d^{r+1}u}{dx^{r+1}}\frac{d^{p-r}v}{dx^{p-r}}\right\} + \cdots +$$

$$+ \left\{\frac{d^p u}{dx^p}\frac{dv}{dx} + v\frac{d^{p+1}u}{dx^{p+1}}\right\}$$

$$= u\frac{d^{p+1}v}{dx^{p+1}} + (1 + {}^pC_1)\frac{du}{dx}\frac{d^p v}{dx^p} + \cdots +$$

$$+ ({}^pC_{r-1} + {}^pC_r)\frac{d^r u}{dx^r}\frac{d^{p-r+1}v}{dx^{p-r+1}} + \cdots + v\frac{d^{p+1}u}{dx^{p+1}}$$

$$= u\frac{d^{p+1}v}{dx^{p+1}} + {}^{p+1}C_1\frac{du}{dx}\frac{d^p v}{dx^p} + \cdots +$$

$$+ {}^{p+1}C_r\frac{d^r u}{dx^r}\frac{d^{p+1-r}v}{dx^{p+1-r}} + \cdots + v\frac{d^{p+1}u}{dx^{p+1}}.$$

Hence, if the theorem is true for $n = p$, then it is also true for $n = p+1$. But the theorem is true for $n = 1$ and, therefore, for all positive integral values of n.

Example.

$$\frac{d^4}{dx^4}(e^{2x}\sin x) = e^{2x}(\sin x) + 4(2e^{2x})(-\cos x) + 6(4e^{2x})(-\sin x) +$$

$$+ 4(8e^{2x})(\cos x) + (16e^{2x})(\sin x)$$
$$= e^{2x}(24\cos x - 7\sin x).$$

Example.

$$\frac{d^n}{dx^n}\left(\frac{1}{x(1+x)}\right) = \frac{1}{x}\frac{(-1)^n\,n!}{(1+x)^{n+1}} + (^nC_1)\left(\frac{-1}{x^2}\right)\frac{(-1)^{n-1}(n-1)!}{(1+x)^n} +$$

$$+ \cdots + \frac{1}{(1+x)}\frac{(-1)^n\,n!}{x^{n+1}}$$

$$= \frac{(-1)^n\,n!}{x^{n+1}(1+x)^{n+1}}\{x^n + x^{n-1}(1+x) + \cdots + (1+x)^n\}$$

$$= \frac{(-1)^n\,n!}{x^{n+1}(1+x)^{n+1}}\{(1+x)^{n+1} - x^{n+1}\},$$

on summing the geometric series,

$$= (-1)^n\,n!\left\{\frac{1}{x^{n+1}} - \frac{1}{(1+x)^{n+1}}\right\}.$$

This result could be obtained more simply by first expanding the function as

$$\frac{1}{x} - \frac{1}{x+1}.$$

Exercises 2(c) (cont.)

2. Find the *n*th differential coefficients of $x^3 e^x$, $x \cos 2x$.

3. If $y = (2x-\pi)^4 \sin\left(\frac{x}{3}\right)$, find the value of $\frac{d^n y}{dx^n}$ when $x = \frac{\pi}{2}$.

4. Find the values of *a, b, c* so that

$$\frac{d^n}{dx^n}\{e^{px}(x^3 + ax^2 + bx + c)\} = e^{px} p^n x^3.$$

5. Differentiate the product $(x^n)(x^n)$ *n* times using Leibniz's theorem. Hence prove that

$$1 + (^nC_1)^2 + (^nC_2)^2 + \cdots + (^nC_n)^2 = \frac{(2n)!}{(n!)^2}.$$

6. Apply Leibniz's theorem to find an expression for $\frac{d^4}{dx^4}(\sin x \sin 3x)$.

By expanding $\sin x \sin 3x$ as $\tfrac{1}{2}(\cos 2x - \cos 4x)$ find the 4th differential coefficient in another form. Identify the two results obtained.

7. Differentiate $\frac{1}{1-x^2}$ *m* times in two ways, (i) by treating the function

as the product $\left(\dfrac{1}{1+x}\right)\left(\dfrac{1}{1-x}\right)$ and applying Leibniz's theorem (ii) by expanding the function in partial fractions and then differentiating. Identify the results obtained by these two methods.

8. We can express the result for the differential coefficient of a product uv, where u and v are functions of x, in terms of operators. Let D denote the operator $\dfrac{d}{dx}$, D_1 the operator which differentiates u and functions of u but leaves v and functions of v untouched, D_2 the operator which differentiates v and functions of v but leaves u and functions of u untouched. Then

$$D(uv) = u\frac{dv}{dx} + v\frac{du}{dx},$$

$$D_1(uv) = v\frac{du}{dx},$$

$$D_2(uv) = u\frac{dv}{dx}.$$

Hence
$$D(uv) = (D_1 + D_2)(uv).$$

More generally

$$(D_1 + D_2)\phi(u)f(v) = f(v)\left[\phi'(u)\frac{du}{dx}\right] + \phi(u)\left[f'(v)\frac{dv}{dx}\right]$$
$$= D[\phi(u)f(v)].$$

Thus we can regard the operator D as equivalent to the operator $D_1 + D_2$ when operating on the product of a function of u and a function of v.

We infer that $D^n(uv) = (D_1 + D_2)^n(uv)$. Assume that you can expand $(D_1 + D_2)^n$ using the Binomial theorem, and hence derive an alternative proof of Leibniz's theorem.

9. If $f(x) = x^2 f^{(2)}(x)$, find $f^{(n)}(x)$ and $f^{(n)}(0)$.

10. If $f(x) = x^4 e^{3x}$, find $f^{(n)}(0)$. If $f(x) = x^3 f'(x)$, find $f^{(n)}(0)$.

11. If $f(x) + x f^{(2)}(x) = 0$, find, by differentiating this equation n times, a relation between $f^{(n)}(0)$ and $f^{(n+1)}(0)$.

Maclaurin's theorem

If we assume that $f(x)$ can be expanded as an absolutely convergent power series $\sum\limits_{n=0}^{\infty} a_n x^n$, we can find the coefficients a_n by repeated differentiation;† on putting $x = 0$ at each stage we can isolate and

† See Page, A. (1947). *Algebra*. p. 195. ULP.

thus evaluate the coefficients a_n. The expansion is

$$f(x) = f(0) + xf^{(1)}(0) + \frac{x^2}{2!}f^{(2)}(0) + \cdots + \frac{x^n}{n!}f^{(n)}(0) + \cdots.$$

In the next chapter we shall discuss conditions for the validity of the expansion but before doing so we shall show the value of the theorem of Leibniz in finding the coefficients in a Maclaurin expansion.

In the remainder of this chapter we are concerned with the application of Maclaurin's theorem to obtain formal expansions without reference to the validity of the process.

Example. $f(x) = \log(1+x)$.

$$f(0) = 0 \quad \text{and} \quad f^{(n)}(x) = \frac{(-1)^{n-1}(n-1)!}{(1+x)^n},$$

whence

$$f^{(n)}(0) = (-1)^{n-1}(n-1)!$$

Hence,

$$\log(1+x) = x - \frac{x^2}{2} + \cdots + (-1)^{n-1}\frac{x^n}{n} + \cdots.$$

This expansion is valid in the range $-1 < x \leq 1$.

Example. $f(x) = e^x \sin x$. $f(0) = 0$ and

$$f^{(n)}(x) = 2^{\frac{1}{2}n} e^x \sin\left(x + \frac{n\pi}{4}\right),$$

whence

$$f^{(n)}(0) = 2^{\frac{1}{2}n} \sin \frac{n\pi}{4}.$$

Hence,

$$e^x \sin x = \sum_{n=1}^{\infty} 2^{\frac{1}{2}n} \sin \frac{n\pi}{4} \frac{x^n}{n!}.$$

This expansion is valid for all values of x.

We could also obtain the expansion by using the fact that $i f(x)$ is the imaginary part of

$$e^x . e^{ix} = e^{(1+i)x} = \sum_{n=0}^{\infty} \frac{(1+i)^n x^n}{n!} = \sum_{n=0}^{\infty} 2^{\frac{1}{2}n} \left(\cos \frac{n\pi}{4} + i \sin \frac{n\pi}{4}\right) \frac{x^n}{n!}$$

whence the result follows.

The first few terms of the expansion may also be found by multiplying the series for e^x and $\sin x$, namely,

$$\left(1+x+\frac{x^2}{2}+\frac{x^3}{6}+\cdots\right)\left(x-\frac{x^3}{6}+\cdots\right),$$

this operation is permissible since both series are absolutely convergent.

Exercises 2(c) (cont.)

12. Use Maclaurin's theorem to find the power series expansion of (i) $\sin x$ (ii) $\cos x$ (iii) $\log(1-x)$.

13. Find, using Maclaurin's theorem, power series expansions as far as the x^5 term of the following functions

    (i) $\tan x$ (ii) $\sin^3 x$ (iii) $e^{\sin x}$.

Check your results by expanding the functions by some other method.

Example.
$$f(x) = e^{a\,\sin^{-1} x}.$$

In this case we cannot find $f^{(n)}(x)$ for a general value of n. Nevertheless, we are able to find the general term of the Maclaurin expansion, as we shall now show.

Let us write $f(x) = y$, so $y = e^{a\,\sin^{-1} x}$. Then

$$\frac{dy}{dx} = e^{a\,\sin^{-1} x}\frac{a}{\sqrt{(1-x)^2}},$$

which may be written as $\sqrt{(1-x^2)}\frac{dy}{dx} = ay$. Differentiating this equation,

$$\sqrt{(1-x^2)}\frac{d^2y}{dx^2} - \frac{x}{\sqrt{(1-x^2)}}\frac{dy}{dx} = a\frac{dy}{dx},$$

$$\therefore (1-x^2)\frac{d^2y}{dx^2} - x\frac{dy}{dx} = a\sqrt{(1-x^2)}\frac{dy}{dx}.$$

Hence
$$(1-x^2)\frac{d^2y}{dx^2} - x\frac{dy}{dx} - a^2 y = 0.$$

On differentiating this equation n times using Leibniz's theorem we obtain

$$\left\{(1-x^2)\frac{d^{n+2}y}{dx^{n+2}}+(n)(-2x)\frac{d^{n+1}y}{dx^{n+1}}+\left(n\cdot\frac{n-1}{2}\right)(-2)\frac{d^ny}{dx^n}\right\}-$$

$$-\left\{x\frac{d^{n+1}y}{dx^{n+1}}+(n)(1)\frac{d^ny}{dx^n}\right\}-a^2\frac{d^ny}{dx^n}=0.$$

Writing $x = 0$ in the last equation, we have

$$\left[\frac{d^{n+2}y}{dx^{n+2}}\right]_{x=0}=(a^2+n^2)\left[\frac{d^ny}{dx^n}\right]_{x=0},$$

or,
$$f^{(n+2)}(0)=(a^2+n^2)f^{(n)}(0).$$

Now $f(0) = 1, f^{(1)}(0) = a, f^{(2)}(0) = a^2$; so, applying the reduction formula,
$$f^{(3)}(0)=(a^2+1^2)a$$

$$f^{(4)}(0)=(a^2+2^2)a^2$$

$$f^{(5)}(0)=(a^2+3^2)(a^2+1^2)a$$

and so on. Hence

$$e^{a\sin^{-1}x}=1+ax+\frac{a^2x^2}{2!}+\frac{a(a^2+1^2)}{3!}x^3+\cdots.$$

Exercises 2(d)

1. If $y = \dfrac{\sin^{-1}x}{\sqrt{(1-x^2)}}$, prove that

$$(1-x^2)\frac{dy}{dx}-xy=1.$$

Differentiate this equation n times using Leibniz's theorem, and show that, if $n \geq 1$,
$$\left[\frac{d^{n+1}y}{dx^{n+1}}\right]_{x=0}=n^2\left[\frac{d^ny}{dx^n}\right]_{x=0}.$$

Hence, using Maclaurin's theorem, give y as a power series in x. Also, obtain the expansion as far as the x^5 term by direct multiplication of the series for $\sin^{-1}x$ and $(1-x^2)^{-\frac{1}{2}}$.

2. If $y = \{x+\sqrt{(1+x^2)}\}^m$, prove that

$$(1+x^2)\frac{d^2y}{dx^2}+x\frac{dy}{dx}-m^2y=0.$$

Differentiate this equation n times using Leibniz's theorem. Then by using Maclaurin's theorem, obtain y as a power series in x.

Show that, when $x = i \sin \theta$, $y = \cos m\theta + i \sin m\theta$. Hence show that

$$\cos m\theta = 1 - \frac{m^2}{2!} \sin^2 \theta + \frac{m^2(m^2-2^2)}{4!} \sin^4 \theta - \frac{m^2(m^2-2^2)(m^2-4^2)}{6!} \sin^6 \theta + \cdots,$$

and write down the corresponding series obtained for $\sin m\theta$.

3. If $y = \tfrac{1}{2}(\sin^{-1} x)^2$, prove that

$$(1-x^2)\frac{d^2y}{dx^2} - x\frac{dy}{dx} - 1 = 0.$$

Prove that

$$\tfrac{1}{2}(\sin^{-1} x)^2 = \sum_{n=1}^{\infty} \frac{2^{2n-2}\{(n-1)!\}^2}{(2n)!} x^{2n}.$$

4. If $x = \sin \theta$, $y = \sinh 2\theta$, when $-\tfrac{1}{2}\pi < \theta < \tfrac{1}{2}\pi$, prove that

$$(1-x^2)\frac{d^2y}{dx^2} - x\frac{dy}{dx} - 4y = 0.$$

Differentiate this equation n times using Leibniz's theorem, and prove that

$$\left[\frac{d^{n+2}y}{dx^{n+2}}\right]_{x=0} = (n^2+4)\left[\frac{d^n y}{dx^n}\right]_{x=0} \quad \text{if} \quad n \geq 0.$$

Deduce the formal expansion of y in terms of x,

$$y = 2\left\{x + \frac{(1^2+4)}{3!}x^3 + \frac{(1^2+4)(3^2+4)}{5!}x^5 + \cdots\right\}.$$

Verify the correctness of this result as far as the x^5 term by expanding y as a power series in θ and then using the expansion of $\theta = \sin^{-1} x$ as a power series in x.

5. If $f(x) = \sin(\sinh^{-1} x)$, prove that

$$f^{(n+2)}(0) = -(n^2+1)f^{(n)}(0).$$

Hence, using Maclaurin's theorem, obtain the expansion of $f(x)$ as a power series in x.

Also obtain the expansion as far as the x^5 term by expanding

$$\sinh^{-1} x = \int_0^x \frac{dt}{\sqrt{(1+t^2)}}$$

as a power series in x as far as the term in x^5, and then using the expansion of $\sin v$ as a power series in v.

6. If $y = \dfrac{2}{3\sqrt{}} \tan^{-1} \dfrac{x\sqrt{3}}{x+2}$, prove that

$$(1+x+x^2)\dfrac{dy}{dx} = 1. \tag{1}$$

If we differentiate this equation n times using Leibniz's theorem and put $x = 0$, we obtain a relation between $\left[\dfrac{d^{n+1}y}{dx^{n+1}}\right]_{x=0}$, $\left[\dfrac{d^n y}{dx^n}\right]_{x=0}$, and $\left[\dfrac{d^{n-1}y}{dx^{n-1}}\right]_{x=0}$. To obtain a simpler relation of this type, we rewrite equation (1) in the form

$$(1-x^3)\dfrac{dy}{dx} = (1-x).$$

Now differentiate this equation n times using Leibniz's theorem and show that

$$\left[\dfrac{d^{n+1}y}{dx^{n+1}}\right]_{x=0} = n(n-1)(n-2)\left[\dfrac{d^{n-2}y}{dx^{n-2}}\right]_{x=0},$$

if $n \geq 2$. Hence, by using Maclaurin's theorem, show that

$$y = x - \dfrac{x^2}{2} + \dfrac{x^4}{4} - \dfrac{x^5}{5} + \dfrac{x^7}{7} \cdots,$$

and give the coefficients of x^{3n}, x^{3n+1}, x^{3n+2} in this series.

Miscellaneous problems on Chapter 2

1. If $f(x) = \sinh(m \sinh^{-1} x)$, prove that

$$f^{(n+2)}(0) = (m^2 - n^2) f^{(n)}(0).$$

Hence, obtain $f(x)$ as a power series in x.
Write down in full the expansion obtained in the case $m = 5$, and verify your result by writing $\sinh^{-1} x = \theta$.

2. If $y = k \sin^{-1} x + (\sin^{-1} x)^2$, where k is a constant, prove that

$$(1-x^2)\dfrac{d^2 y}{dx^2} - x\dfrac{dy}{dx} - 2 = 0.$$

Find the expansion of y as a power series in x, giving the coefficients of the terms in x^{2n} and x^{2n+1}.

3. Prove that $\dfrac{d^n}{dx^n}(x^n \log x) = n!(\log x + a_1 - \tfrac{1}{2}a_2 + \tfrac{1}{3}a_3 - \cdots)$ where a_r is the coefficient of x^r in the expansion of $(1+x)^n$.

4. Prove that when $x = 1$

$$\dfrac{d^n}{dx^n}(e^x x^m) = \dfrac{d^m}{dx^m}(e^x x^n).$$

5. If $x^2 y = \sin^2 x$, prove that

$$x^2 \frac{d^2 y}{dx^2} + 4x \frac{dy}{dx} + 2y(1+2x^2) = 2.$$

Differentiate this equation n times using Leibniz's theorem. Prove that, if n is odd, $\left[\dfrac{d^n y}{dx^n}\right]_{x=0} = 0$ and find the value of $\left[\dfrac{d^n y}{dx^n}\right]_{x=0} \Big/ \left[\dfrac{d^{n-2} y}{dx^{n-2}}\right]_{x=0}$, when n is even.

6. Obtain the coefficients of x^2, x^3, x^4 in the expansion of

$$y = (1-x+x^2)^n$$

in two ways: (i) by straightforward binomial expansion, (ii) by using Maclaurin's theorem.

7. If $f(x) = x^{\frac{1}{2}} \tan^{-1}(x^{\frac{1}{2}})$, prove that

$$(2n+1) f^{(n+1)}(0) + (2n-1)(n+1) f^{(n)}(0) = 0.$$

Obtain the coefficient of x^n in the power series expansion of $f(x)$.

8. From the result

$$\log\{x+\sqrt{(1+x^2)}\} = \int_0^x \frac{dt}{\sqrt{(1+t^2)}},$$

prove that, if $|x| < 1$,

$$\log\{x+\sqrt{(1+x^2)}\} = x - \frac{1^2}{3!} x^3 + \frac{1^2 . 3^2}{5!} x^5 - \frac{1^2 . 3^2 . 5^2}{7!} x^7 + \cdots.$$

Deduce that if $f(x) = \log\{x+\sqrt{(1+x^2)}\}$

and
$$f^{(2n)}(0) = 0,$$
$$f^{(2n+1)}(0) = (-1)^n 1^2 . 3^2 . 5^2 \ldots (2n-1)^2.$$

These results can also be proved by establishing the formula

$$(1+x^2) f''(x) + x f'(x) = 0,$$

differentiating n times using Leibniz's theorem, putting $x = 0$, and then using an induction proof.

9. From the result $\sin^{-1} x = \displaystyle\int_0^x \frac{dt}{\sqrt{(1-t^2)}}$ prove that, if $|x| < 1$,

$$\sin^{-1} x = x + \frac{1^2}{3!} x^3 + \frac{1^2 . 3^2}{5!} x^5 + \cdots.$$

Deduce that if $f(x) = \sin^{-1} x$,
$$f^{(2n)}(0) = 0$$
and
$$f^{(2n+1)}(0) = 1^2 . 3^2 . 5^2 \cdots (2n-1)^2.$$

10. If $y = \tanh^{-1} x$, find $\dfrac{d^n y}{dx^n}$.

Discuss the convergence or divergence of the series $\sum\limits_{n=1}^{\infty} \dfrac{d^n y}{dx^n}$.

11. Use Maclaurin's theorem to establish the following expansions as far as the terms indicated.

(i) $\sec x = 1 + \tfrac{1}{2}x^2 + \tfrac{5}{24}x^4 + \cdots$.

(ii) $\tan x = x + \tfrac{1}{3}x^3 + \tfrac{2}{15}x^5 + \cdots$.

(iii) $\dfrac{x^2}{x - \log(1+x)} = 2 + \tfrac{4}{3}x - \tfrac{1}{9}x^2 + \cdots$.

These results may be obtained more simply by operating on the well-known series for $\cos x$, $\sin x$, $\log(1+x)$. See Page, A. (1947) *Algebra*, pp. 212, 213. ULP.

12. If
$$f(x) = \frac{\{x + \sqrt{(1+x^2)}\}^m}{\sqrt{(1+x^2)}},$$
prove that
$$f^{(n+2)}(0) = \{m^2 - (n+1)^2\} f^{(n)}(0),$$
and hence obtain the Maclaurin expansion of $f(x)$, and also of $f(x) + f(-x)$. In the latter expansion put $x = i \sin \theta$ and prove that

$$\frac{\cos m\theta}{\cos \theta} = 1 - \frac{(m^2 - 1^2)}{2!} \sin^2 \theta + \frac{(m^2 - 1^2)(m^2 - 3^2)}{4!} \sin^4 \theta - \cdots.$$

If m is odd, the series on the right is finite and we have the expansion of $\dfrac{\cos m\theta}{\cos \theta}$ as a polynomial in $\sin \theta$.

By considering the expansion of $f(x) - f(-x)$, obtain the expansion of $\sin m\theta$ as a polynomial in $\sin \theta$.

3

The Mean Value Theorem and Its Applications

Rolle's theorem

Suppose that $f(x)$ is continuous in the closed interval $a \leq x \leq b$ and differentiable in the open† interval $a < x < b$. Then if $f(a) = 0$ and $f(b) = 0$, there is at least one value of x between a and b for which $f'(x) = 0$.

For either $f(x)$ is constant between $x = a$ and $x = b$ in which case the theorem is trivial, or from the graph of $y = f(x)$ there is at least one value of x in the range $a < x < b$ for which $f(x)$ has either a maximum or minimum value.‡ If x_1 is the value of x at any one of these maximum or minimum points, then $f'(x_1) = 0$, which is the theorem.

In Fig. 3.1, there is precisely one value of x in the range $a < x < b$ for which $f'(x) = 0$; in Fig. 3.2, there are two such values of x in the range. In Fig. 3.3, there are four such values of x in the range.

It is clearly necessary for the function $f(x)$ to be differentiable at all points of the range $a < x < b$; e.g. the function whose graph is shown in Fig. 3.4 is not differentiable at P and there is no value of x in the range $a < x < b$ for which $f'(x) = 0$.

The condition that $f(x)$ must be continuous in the *closed* interval $a \leq x \leq b$ is clearly necessary; for consider the function defined by the equations
$$f(x) = x \quad \text{if} \quad 1 < x < 2,$$
$$f(1) = 0,$$
$$f(2) = 0.$$

---

† The closed interval $a \leq x \leq b$ is usually denoted by $[a, b]$ and the open interval $a < x < b$ by (a, b).

‡ It is possible to avoid recourse to graphical argument by proving the theorem that a function continuous in a closed interval attains its upper and lower bounds.

This function is continuous for $1 < x < 2$, but not at the end points $x = 1$, $x = 2$. The function is *not* continuous in the closed interval $1 \le x \le 2$, and hence Rolle's theorem does not apply. It

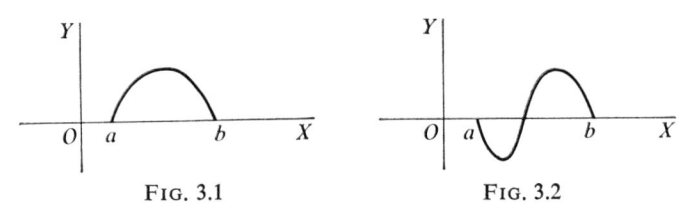

FIG. 3.1

FIG. 3.2

is clear from the graph of the function in Fig. 3.5 that there is no value of x in the range $1 < x < 2$ for which $f'(x) = 0$.

On the other hand, it is not necessary for the function to be differentiable at the end points of the range. This is illustrated by

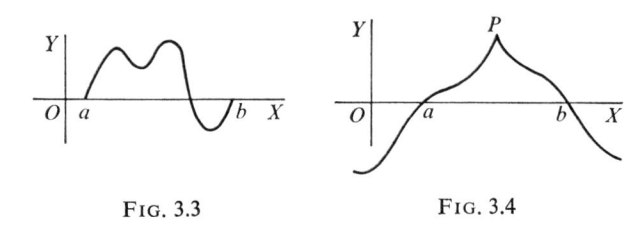

FIG. 3.3

FIG. 3.4

the function whose graph is shown in Fig. 3.6. The function is continuous for $1 \le x \le 2$ and differentiable for $1 < x < 2$, but is not differentiable at the end points $x = 1$, $x = 2$. But there is a value of x in the range $1 < x < 2$ for which $f'(x) = 0$.

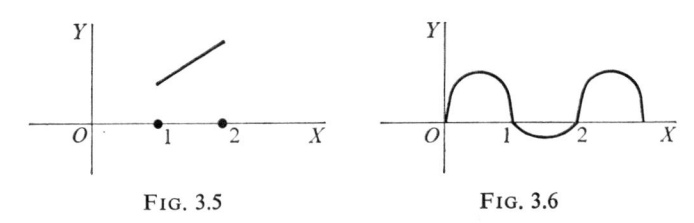

FIG. 3.5

FIG. 3.6

An alternative statement of Rolle's theorem is as follows: if $f(x)$ is continuous in the range $a \le x \le b$ and differentiable for $a < x < b$, then between any two roots of the equation $f(x) = 0$, there is at least one root of the equation $f'(x) = 0$.

Exercises 3(a)

1. Prove the slightly more general form of Rolle's theorem:
If $f(x)$ is continuous in the closed interval $a \leq x \leq b$ and differentiable in the open interval $a < x < b$ and $f(a) = f(b)$, then there is at least one value of x between a and b for which $f'(x) = 0$.

2. State whether the generalized Rolle's theorem applies to the following functions:

(i) $y = \sin x$, $a = 0$, $b = \pi$.
(ii) $y = x^{-2}$, $a = -1$, $b = 1$.
(iii) $y = \dfrac{(x-1)(x-3)}{(x-2)}$, $a = 1$, $b = 3$.
(iv) $y = x^4$, $a = -2$, $b = 2$.
(v) $y = \tan x$, $a = \pi$, $b = 2\pi$.

Rolle's theorem is the basis of many interesting and important results in analysis. We shall first apply the theorem to the problem of locating the roots of an equation.

Example. If $f(x) = (x-a)(x-b)(x-c)$, where $a < b < c$, then $f(a) = 0$, $f(b) = 0$, $f(c) = 0$ and

$$f'(x) = (x-a)(x-b) + (x-a)(x-c) + (x-b)(x-c).$$

It follows that the quadratic equation $f'(x) = 0$ has two real roots, one between a and b, the other between b and c. Show that the equation has two real roots by the normal method for investigating the roots of a quadratic equation.

Exercises 3(a) (cont.)

3. If $f(x) = (x-1)(x-2)(x-3)(x-4)$, give some information about the roots of the equation $f'(x) = 0$. Check your results by finding the roots of this cubic equation to one decimal place.

4. If $f(x)$ is a polynomial of degree n and all the roots of $f(x) = 0$ are real, prove that all the roots of $f'(x) = 0$ are real. Prove also that all the roots of $f''(x) = 0$ are real. What can you say about the roots of $f^{(r)}(x) = 0$?

Multiple roots of an equation

Consider the equation $f(x) = 0$, where $f(x)$ is a polynomial in x. By letting b tend to a in Rolle's theorem we see that if $f(x) = 0$ has a double root $x = a$, then $f'(x) = 0$ also has a root $x = a$.

This important test of the existence of double roots of an equation can be proved quite simply without recourse to Rolle's theorem.

For if $f(x) = 0$ has a double root $x = a$, we may write
$$f(x) = (x-a)^2\phi(x),$$
where $\phi(x)$ is a polynomial in x. Then
$$f'(x) = (x-a)^2\phi'(x) + 2(x-a)\phi(x),$$
whence $f'(a) = 0$.

If $f(x) = 0$ has a threefold root $x = a$, we may write $f(x)$ in the form
$$f(x) = (x-a)^3\phi(x),$$
whence
$$f'(x) = (x-a)^3\phi'(x) + 3(x-a)^2\phi(x)$$
$$= (x-a)^2 F(x), \text{ say.}$$

Hence, $f'(x) = 0$ has a double root $x = a$ and therefore $f''(x) = 0$ has a single root $x = a$. Thus, a threefold root of $f(x) = 0$ is also a root of $f'(x) = 0$ and $f''(x) = 0$.

In general, if $f(x) = 0$ has an r-fold root $x = a$, then $x = a$ is also a root of the equations
$$f'(x) = 0, f''(x) = 0, \ldots f^{(r-1)}(x) = 0.$$

Example. The equation $f(x) = 0$, where $f(x) = x^2(2-x)^3$, has roots $x = 0$ twice, $x = 2$ three times. The equation $f'(x) = 0$ of degree 4 has roots $x = 0$ single, $x = \xi_1$ between 0 and 2, $x = 2$ twice. The equation $f''(x) = 0$ of degree 3 has roots, $x = \xi_2$ between 0 and ξ_1, $x = \xi_3$ between ξ_1 and 2, $x = 2$ single. The equation $f'''(x) = 0$ of degree 2 has roots $x = \xi_4$ between ξ_2 and ξ_3, $x = \xi_5$ between ξ_3 and 2. Thus $f'''(x) = 0$ has two real roots, each lying between 0 and 2. In fact, the equation $f'''(x) = 0$ is
$$-60x^2 + 144x - 72 = 0$$
and the roots are 1·69 and 0·71.

Example. Find the values of a, b, c in order that the equation
$$x^5 + ax^2 + bx + c = 0$$
should have a treble root $x = 1$.

If $x = 1$ is a treble root of the equation, it must satisfy
$$x^5 + ax^2 + bx + c = 0,$$
$$5x^4 + 2ax + b = 0,$$
$$20x^3 + 2a = 0.$$

Hence
$$1+a+b+c = 0,$$
$$5+2a+b = 0,$$
$$20+2a = 0.$$
$$\therefore \quad a = -10, \quad b = 15, \quad c = -6.$$

The result could be obtained otherwise by equating coefficients in the identity
$$x^5+ax^2+bx+c \equiv (x-1)^3(x^2+px+q).$$

Exercises 3(a) (cont.)

5. Give some information about the roots of the equation
$$f'''(x) = 0 \quad \text{when} \quad f(x) = x^4(1-x)^3.$$

6. Give some information about the roots of the equation $f^{(4)}(x) = 0$ when $f(x) = (x-1)^8(x-2)^7$.

7. Find the condition that the equation

(i) $x^3+ax+b = 0$ (ii) $x^n+ax+b = 0$

should have a pair of equal roots.

8. Find the conditions that the equation $x_u+ax^2+bx+c = 0$ should have a treble root.

9. Find the values of a, b, c in order that the equation
$$x^4+ax^2+bx+c = 0$$
should have a double root $x = 2$ and a single root $x = -1$.

Example. Investigate whether the equation
$$f(x) \equiv 8x^4+20x^3-2x^2-17x+6 = 0 \tag{1}$$
has any multiple roots.

Any multiple root $x = \alpha$ of the equation $f(x) = 0$ satisfies the equations $f(x) = 0$ and $f'(x) = 0$. Thus $x = \alpha$ is a common root of (1) and
$$32x^3+60x^2-4x-17 = 0. \tag{2}$$

There are various techniques for finding a common root of a pair of equations. On multiplying eqn (2) by x and then eliminating the x^4 term using eqn (1) we see that $x = \alpha$ satisfies
$$20x^3-4x^2-51x+24 = 0. \tag{3}$$

THE MEAN VALUE THEOREM AND ITS APPLICATIONS 63

On eliminating the x^3 term from eqn (2) and eqn (3), $x = \alpha$ satisfies

$$332x^2+388x-277 = 0. \tag{4}$$

From eqn (3) and eqn (4), $x = \alpha$ satisfies

$$284x^2+356x-249 = 0. \tag{5}$$

Eliminating x^2 from eqn (4) and eqn (5), $x = \frac{1}{2}$. It is easily seen that $f(\frac{1}{2}) = 0$, $f'(\frac{1}{2}) = 0$, $f''(\frac{1}{2}) \neq 0$. We deduce that $x = \frac{1}{2}$ is a double root of the equation $f(x) = 0$.

This result could have been found more easily by factorizing $f(x)$ or by using standard formulae for the sum, etc. of roots of an equation. The other roots of $f(x) = 0$ are $x = -2, -\frac{3}{2}$.

Another technique for finding a common root of a pair of equations is as follows. Suppose $f(x) = 0$ and $\phi(x) = 0$ have a common root $x = \alpha$, and the degree of $f(x)$ is greater than that of $\phi(x)$. Then, if we divide $f(x)$ by $\phi(x)$ and obtain a remainder $g(x)$ so that

$$f(x) = \phi(x)q(x)+g(x),$$

then clearly $g(x) = 0$ also has a root $x = \alpha$, and the degree of $g(x)$ is less than that of $\phi(x)$. If now we divide $\phi(x)$ by $g(x)$ and obtain a remainder $h(x)$ so that

$$\phi(x) = g(x)q_1(x)+r(x),$$

then clearly $r(x) = 0$ also has a root $x = \alpha$, and the degree of $r(x)$ is less than that of $g(x)$. We then divide $g(x)$ by $r(x)$, etc. Proceeding in this way until we get a zero remainder we can isolate the root α.

Exercises 3(a) (cont.)

10. Find the multiple roots of the equation

$$x^4-6x^3+x^2+24x+16 = 0.$$

11. Prove that if $f(x)$ is a polynomial then any root which occurs p times in $f(x) = 0$ will occur $p-q$ times in $f^{(q)}(x) = 0$.

12. Show that the equation

$$x^4+7x^3+12x^2-4x-16 = 0$$

has a double root $x = -2$.

The first mean value theorem

If $f(x)$ is continuous in the range $a \leq x \leq b$, and differentiable in the range $a < x < b$, then, by considering the graph of $f(x)$ between

$x = a$ and $x = b$ (see Fig. 3.7), it is geometrically intuitive that there is at least one point R between P and Q for which the tangent to the curve at R is parallel to the chord PQ. Thus, if the x-coordinate of R is ξ,

$$\frac{f(b)-f(a)}{b-a} = f'(\xi),$$

or $f(b)-f(a) = (b-a)f'(\xi)$, where $a < \xi < b$. This result is known as the First Mean Value theorem. We shall give a formal proof of the theorem using Rolle's theorem.

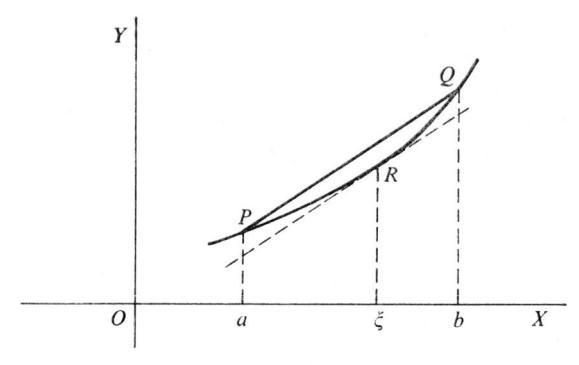

Fig. 3.7

Consider the function

$$\phi(x) = (x-a)[f(b)-f(a)] - (b-a)[f(x)-f(a)].$$

It is clear that $\phi(x)$ is continuous in $[a, b]$ and differentiable in (a, b), and also that $\phi(a) = 0$, $\phi(b) = 0$. Hence, by Rolle's theorem, $\phi'(x)$ vanishes at an inner point of the range.

Now
$$\phi'(x) = f(b)-f(a)-(b-a)f'(x),$$

and hence there exists a number ξ such that

$$f(b)-f(a)-(b-a)f'(\xi) = 0,$$

where $a < \xi < b$, i.e. $f(b)-f(a) = (b-a)f'(\xi)$. Putting $b = a+h$, we can write ξ in the form $a+\theta h$, where $0 < \theta < 1$, and the mean value theorem takes the form

$$f(a+h)-f(a) = hf'(a+\theta h),$$

where $0 < \theta < 1$. If h is small and $f'(x)$ is continuous, $f'(a+\theta h)$ is approximately equal to $f'(a)$ whence $f(a+h)-f(a) = hf'(a)$ approximately, if h is small, a result we have used previously.

Example. Taking $f(x) = \sin x$, we deduce that
$$\sin b - \sin a = (b-a)\cos \xi,$$
where $a < \xi < b$.

Taking $f(x) = e^x$, we deduce that
$$e^b - e^a = (b-a)e^\xi,$$
where $a < \xi < b$.

Exercises 3(b)

1. Find the precise value of ξ in the following cases:
   (i) $f(x) = x^3$, $a = 1$, $b = 2$.
   (ii) $f(x) = x^4$, $a = 0$, $b = 3$.

2. Prove that if $f(x)$ and $\phi(x)$ are continuous in $[a, b]$ and differentiable in (a, b), there exists a number ξ such that
$$f(a)\phi(b) - \phi(a)f(b) = (b-a)\{f(a)\phi'(\xi) - \phi(a)f'(\xi)\},$$
where $a < \xi < b$.
Find the precise value of ξ in the case $f(x) = x^2$, $\phi(x) = x^3$, $a = 1$, $b = 2$.

3. Use the first mean value theorem to prove that, if c is any value in the range (a, b),
$$\lim_{x \to c} \frac{f(x)-f(c)}{\phi(x)-\phi(c)} = \frac{f'(c)}{\phi'(c)},$$
provided $f'(x)$ and $\phi'(x)$ are continuous in (a, b) and $\phi'(c) \neq 0$. It follows that, if $f(c) = 0$ and $\phi(c) = 0$,
$$\lim_{x \to c} \frac{f(x)}{\phi(x)} = \frac{f'(c)}{\phi'(c)}.$$
Use this result to evaluate the following limits:
$$\lim_{x \to 1} \frac{e^x - e}{x-1}\,;\quad \lim_{x \to 2} \frac{\log(\tfrac{1}{2}x)}{2-x}\,;\quad \lim_{x \to \tfrac{1}{2}\pi} \frac{1-\sin x}{\cos x}.$$

4. $f'(x)$ is continuous in $[a, b]$ and differentiable in (a, b). The function $\phi(x)$ is defined by
$$\phi(x) = -f(b)+f(x)+(b-x)f'(x)+\tfrac{1}{2}(b-x)^2 R,$$
where R is a constant chosen to make $\phi(a) = 0$. Apply Rolle's theorem to $\phi(x)$ and deduce that $R = f''(\xi)$, where $a < \xi < b$.

It follows that
$$f(b) = f(a)+(b-a)f'(a)+\tfrac{1}{2}(b-a)^2 f''(\xi),$$
where $a < \xi < b$, a result known as the second mean value theorem.

Find the precise values of ξ in the case $f(x) = x^4$, $a = -1$, $b = 2$.

The general mean value theorem

We assume that $f(x)$ is differentiable up to the nth order, i.e. that $f^{(n)}(x)$ exists, in the range $a < x < b$, and that $f^{(n-1)}(x)$ is continuous in the range $a \leq x \leq b$. Consider the function

$$\phi(x) = -f(b)+f(x)+(b-x)f'(x)+\frac{(b-x)^2}{2!}f''(x)+$$
$$+\cdots+\frac{(b-x)^{n-1}}{(n-1)!}f^{(n-1)}(x)+\frac{(b-x)^n}{n!}R,$$

where R is a constant, independent of x, chosen to make $\phi(a) = 0$, i.e.

$$0 = -f(b)+f(a)+(b-a)f'(a)+$$
$$+\cdots+\frac{(b-a)^{n-1}}{(n-1)!}f^{(n-1)}(a)+\frac{(b-a)^n}{n!}R. \quad (1)$$

Clearly $\phi(x)$ is continuous in $[a, b]$ and differentiable in (a, b), and also $\phi(a) = 0$, $\phi(b) = 0$. Hence, by Rolle's theorem, $\phi'(x)$ must vanish at some point of the range (a, b).

Now

$$\phi'(x) = f'(x)+\{(b-x)f''(x)-f'(x)\}+$$
$$+\left\{\frac{(b-x)^2}{2!}f^{(3)}(x)-(b-x)f''(x)\right\}+$$
$$+\left\{\frac{(b-x)^3}{3!}f^{(4)}(x)-\frac{(b-x)^2}{2!}f^{(3)}(x)\right\}+$$
$$+\cdots+\left\{\frac{(b-x)^{n-1}}{(n-1)!}f^{(n)}(x)-\frac{(b-x)^{n-2}}{(n-2)!}f^{(n-1)}(x)\right\}-$$
$$-\frac{(b-x)^{n-1}}{(n-1)!}R$$
$$= \frac{(b-x)^{n-1}}{(n-1)!}\{f^{(n)}(x)-R\}.$$

THE MEAN VALUE THEOREM AND ITS APPLICATIONS 67

Hence $f^{(n)}(x) - R = 0$ at some point ξ of the range (a, b), i.e. $R = f^{(n)}(\xi)$, where $a < \xi < b$.

Substituting this value of R in eqn (1) we have

$$f(b) = f(a) + (b-a)f'(a) + \frac{(b-a)^2}{2!}f''(a) +$$
$$+ \cdots + \frac{(b-a)^{n-1}}{(n-1)!}f^{(n-1)}(a) + \frac{(b-a)^n}{n!}f^{(n)}(\xi),$$

where $a < \xi < b$. We can write ξ in the form $a + \theta(b-a)$, where $0 < \theta < 1$. On putting $b = a + x$, the theorem takes the form

$$f(a+x) = f(a) + xf'(a) + \frac{x^2}{2!}f''(a) +$$
$$+ \cdots + \frac{x^{n-1}}{(n-1)!}f^{(n-1)}(a) + \frac{x^n}{n!}f^{(n)}(a+\theta x),$$

where $0 < \theta < 1$. This result is known as Taylor's theorem. On putting $b = x$, $a = 0$, the theorem takes the form

$$f(x) = f(0) + xf'(0) + \frac{x^2}{2!}f''(0) + \cdots + \frac{x^{n-1}}{(n-1)!}f^{(n-1)}(0) + \frac{x^n}{n!}f^{(n)}(\theta x),$$

where $0 < \theta < 1$.

The Maclaurin expansion of a function

The last result can be used to determine conditions for the validity of the Maclaurin expansion of a function $f(x)$, which we assume is differentiable up to all orders. For, if S_n denotes the sum of n terms of the infinite series

$$f(0) + xf'(0) + \frac{x^2}{2!}f''(0) + \cdots + \frac{x^n}{n!}f^{(n)}(0) + \cdots,$$

it follows from the general mean value theorem that

$$f(x) = S_n + \frac{x^n}{n!}f^{(n)}(\theta x),$$

or

$$S_n = f(x) - \frac{x^n}{n!}f^{(n)}(\theta x).$$

If we can prove that $\frac{x^n}{n!}f^{(n)}(\theta x)$ tends to zero as n tends to infinity, it will follow that S_n must tend to the limit $f(x)$ as n tends to infinity. Thus the infinite series will converge to a sum $f(x)$, i.e.

$$f(x) = f(0) + xf'(0) + \frac{x^2}{2!}f''(0) + \cdots,$$

provided that $\lim\limits_{n \to \infty} \frac{x^n}{n!} f^{(n)}(\theta x) = 0$.

Power series for cos x

If $f(x) = \cos x$, $f^{(n)}(x) = \cos\left(x + \frac{n\pi}{2}\right)$ and

$$\left|\frac{x^n}{n!} f^{(n)}(\theta x)\right| = \left|\frac{x^n}{n!} \cos\left(\theta x + \frac{n\pi}{2}\right)\right| < \frac{|x|^n}{n!},$$

which tends to zero as n tends to infinity for all values of x. Hence

$$\cos x = 1 - \frac{x^2}{2!} + \frac{x^4}{4!} - \cdots$$

for all values of x. Similarly

$$\sin x = x - \frac{x^3}{3!} + \frac{x^5}{5!} - \cdots$$

for all values of x.

Power series for $\log(1+x)$

If

$$f(x) = \log(1+x), \quad f^{(n)}(x) = \frac{(-1)^{n-1}(n-1)!}{(1+x)^n},$$

and

$$\left|\frac{x^n}{n!} f^{(n)}(\theta x)\right| = \left|\frac{x^n}{n!} \frac{(-1)^{n-1}(n-1)!}{(1+\theta x)^n}\right|$$

$$= \frac{1}{n}\left|\frac{x}{1+\theta x}\right|^n.$$

Now θ lies between 0 and 1 and clearly if $-\frac{1}{2} \leq x \leq 1$,

$$\left|\frac{x}{1+\theta x}\right| \leq 1,$$

and then

$$\left|\frac{x^n}{n!} f^{(n)}(\theta x)\right| \leq \frac{1}{n},$$

and tends to zero as n tends to infinity. Hence

$$\log(1+x) = x - \frac{x^2}{2} + \frac{x^3}{3} - \cdots$$

if $-\frac{1}{2} \leq x \leq 1$. Thus the expression $\frac{x^n}{n!} f^{(n)}(\theta x)$ for the remainder†

† This form of the remainder is called Lagrange's form.

after n terms of the Maclaurin expansion is not precise enough to establish the series for $\log(1+x)$ for the full range of validity $-1 < x \leq 1$. To establish the series over the full range it is necessary to estimate the remainder in a different manner and this is now done.

Cauchy's form of the remainder in the Maclaurin series

Consider the function of t,

$$\phi(t) = -f(x) + f(t) + (x-t)f'(t) + \frac{(x-t)^2}{2!} f''(t) +$$
$$+ \cdots + \frac{(x-t)^{n-1}}{(n-1)!} f^{(n-1)}(t).$$

Then

$$\phi'(t) = \frac{(x-t)^{n-1}}{(n-1)!} f^{(n)}(t).$$

Also $\phi(x) = 0$ and

$$\phi(0) = -f(x) + f(0) + xf'(0) + \cdots + \frac{x^{n-1}}{(n-1)!} f^{(n-1)}(0).$$

We apply the first mean value theorem to $\phi(t)$ in the range of values of $t(0, x)$, viz.

$$\phi(x) = \phi(0) + x\phi'(\theta x), \quad \text{where} \quad 0 < \theta < 1.$$

Clearly,

$$\phi'(\theta x) = \frac{(x - \theta x)^{n-1}}{(n-1)!} f^{(n)}(\theta x).$$

Hence

$$0 = \left[-f(x) + f(0) + \cdots + \frac{x^{n-1}}{(n-1)!} f^{(n-1)}(0) \right] +$$
$$+ \frac{x^n}{(n-1)!} (1-\theta)^{n-1} f^{(n)}(\theta x),$$

i.e.
$$f(x) = f(0) + xf'(0) + \cdots + \frac{x^{n-1}}{(n-1)!} f^{(n-1)}(0) +$$
$$+ \frac{x^n}{(n-1)!}(1-\theta)^{n-1} f^{(n)}(\theta x).$$

It follows that the expansion of $f(x)$ as an infinite series
$$f(x) = f(0) + xf'(0) + \frac{x^2}{2!} f''(0) + \cdots$$
is valid if $\frac{x^n}{(n-1)!}(1-\theta)^{n-1} f^{(n)}(\theta x)$ tends to zero as n tends to infinity. This form of the result is due to Cauchy. For example, the logarithmic series,
$$\log(1+x) = x - \frac{x^2}{2} + \frac{x^3}{3} - \cdots$$
is valid provided
$$R = \frac{x^n}{(n-1)!}(1-\theta)^{n-1} \frac{(-1)^{n-1}(n-1)!}{(1+\theta x)^n}$$
tends to zero as n tends to infinity. Now
$$|R| = \frac{|x|^n}{1-\theta} \cdot \left| \frac{1-\theta}{1+\theta x} \right|^n.$$
If $|x| < 1$, $\left|\frac{1-\theta}{1+\theta x}\right| < 1$, and $|R| < \frac{|x|^n}{1-\theta}$ and tends to zero as n tends to infinity. Thus we prove the validity of the logarithmic series for $|x| < 1$, which with the previous result establishes it for $-1 < x \leq 1$.

The binomial theorem

Consider $f(x) = (1+x)^m$, where m is not a positive integer. Then
$$f^{(n)}(x) = m(m-1)(m-2)\cdots(m-n+1)(1+x)^{m-n},$$
and Cauchy's form of the remainder after n terms of the Maclaurin expansion of $f(x)$ is
$$R_n = \frac{x^n}{(n-1)!} \cdot (1-\theta)^{n-1} \cdot m(m-1)(m-2)\cdots(m-n+1)(1+\theta x)^{m-n}.$$

It follows that
$$\frac{R_{n+1}}{R_n} = \frac{x}{n}(1-\theta)(m-n)\frac{1}{1+\theta x}$$
$$= \frac{m-n}{n}\left(\frac{x-\theta x}{1+\theta x}\right).$$

Now, if $|x| < 1$, $\left|\frac{x-\theta x}{1+\theta x}\right| < 1$, and hence $\lim_{n\to\infty}\left|\frac{R_{n+1}}{R_n}\right| < 1$. Hence, by D'Alembert's test, the series $\Sigma |R_n|$ converges, and hence $R_n \to 0$ as n tends to infinity. It follows that the Maclaurin expansion of $f(x)$ is valid if $|x| < 1$, i.e.

$$(1+x)^m = 1+mx+\frac{m(m-1)}{2!}x^2+\cdots,$$

if $|x| < 1$.

A simpler proof of the binomial theorem is given by Page in *Algebra*, p. 155.

It is always necessary to prove that the remainder term, in one of the forms $\frac{x^n}{n!}f^{(n)}(\theta x)$ or $\frac{x^n}{(n-1)!}(1-\theta)^{n-1}f^{(n)}(\theta x)$, tends to zero as n tends to infinity in order to establish the validity of the Maclaurin expansion of $f(x)$. This is by no means an easy task since, except for relatively simple functions, it is not possible to find a formula for $f^{(n)}(x)$. But the reader who goes on to study functions of a complex variable in later chapters will find that the difficulties inherent in the Maclaurin expansion of a function of a real variable melt away when we extend the domain of the variable to complex numbers.

It might be expected that if we prove that the series

$$f(0)+xf'(0)+\frac{x^2}{2!}f''(0)+\cdots$$

converges, this would be sufficient to establish the fact that its sum is $f(x)$. But, unfortunately, this is not true as Cauchy pointed out by considering the function defined by

$$f(x) = e^x - e^{-1/x^2} \quad \text{for} \quad x \neq 0,$$
$$= 1 \quad \text{for} \quad x = 0.$$

This function is continuous and differentiable for all values of x. Clearly, $f(0) = 1, f'(0) = 1, f''(0) = 1$, etc. Thus the series

$$f(0)+xf'(0)+\cdots$$

is

$$1+x+\frac{x^2}{2!}+\cdots = e^x,$$

but $f(x) = e^x - e^{-1/x^2}$ if $x \neq 0$, so the Maclaurin expansion converges but not to the value of the function.

Applications of Taylor's theorem

We shall now use Taylor's theorem in the form

$$f(a+h) = f(a)+hf'(a)+\cdots+\frac{h^{n-1}}{(n-1)!}f^{(n-1)}(a)+\frac{h^n}{n!}f^{(n)}(a+\theta h)$$

in the investigation of a number of problems beginning with a consideration of maximum and minimum values.

Maximum and minimum values

If $f(a+h)-f(a)$ has the same sign for all sufficiently small values of h, then clearly $f'(a)$ must be zero. Then

$$f(a+h)-f(a) = \frac{h^2}{2!}f''(a)+\cdots.$$

For small values of h, the sign of the right-hand side is dominated by the term $\frac{h^2}{2!}f''(a)$ provided $f''(a) \neq 0$. If $f''(a)$ is positive $f(a+h)-f(a)$ is positive and then $x = a$ gives a minimum value of $f(x)$; if $f''(a)$ is negative, $f(a+h)-f(a)$ is negative and then $x = a$ gives a maximum value of $f(x)$.

But if $f''(a) = 0$, the sign of the right-hand side is dominated by the term $\frac{h^3}{3!}f^{(3)}(a)$ provided $f^{(3)}(a) \neq 0$; this term changes sign as h goes from negative to positive; in this case we cannot have a maximum or minimum, and $x = a$ is a point of inflexion, but if $f^{(3)}(a) = 0$ the right-hand side is dominated by the term $\frac{h^4}{4!}f^{(4)}(a)$ provided $f^{(4)}(a) \neq 0$. If $f^{(4)}(a)$ is positive $f(a+h)-f(a)$ is positive and then $x = a$ gives a minimum value of $f(x)$. If $f^{(4)}(a)$ is negative, $f(a+h)-f(a)$ is negative and then $x = a$ gives a maximum value

THE MEAN VALUE THEOREM AND ITS APPLICATIONS

of $f(x)$. It will now be clear that if the first derivative not to vanish at $x = a$ is $f^{(n)}(a)$, where n is even, then we have a maximum or minimum at $x = a$; if $f^{(n)}(a) < 0$, $f(a)$ is a maximum value; if $f^{(n)}(a) > 0$, $f(a)$ is a minimum value.

If, on the other hand, the first derivative not to vanish at $x = a$ is $f^{(n)}(a)$, where n is odd, then $x = a$ gives a point of inflexion.

Example. If
$$f(x) = (x-1)^5(x-2)^4,$$
$$f'(x) = (x-1)^4(x-2)^3(9x-14),$$
and vanishes when $x = 1, 2, \tfrac{14}{9}$.

Clearly, the first derivative not to vanish when $x = 1$ is the fifth; the first derivative not to vanish when $x = 2$ is the fourth; the first derivative not to vanish when $x = \tfrac{14}{9}$ is the second. Hence $x = 2$ and $x = \tfrac{14}{9}$ give maximum or minimum values of $f(x)$; $x = 1$ gives a point of inflexion. It is easy to see that $x = 2$ gives a minimum value and $x = \tfrac{14}{9}$ a maximum value.

L'Hôpital's rules for limits

We consider the limiting value as x tends to a of $\dfrac{f(x)}{\phi(x)}$, where $f(a) = 0$ and $\phi(a) = 0$. Let us assume that $f'(x)$ and $\phi'(x)$ exist and are continuous in the neighbourhood of $x = a$. Then

$$\lim_{x \to a} \frac{f(x)}{\phi(x)} = \lim_{h \to 0} \frac{f(a+h)}{\phi(a+h)} = \lim_{h \to 0} \frac{f(a) + hf'(a+\theta_1 h)}{\phi(a) + h\phi'(a+\theta_2 h)}$$

$$= \lim_{h \to 0} \frac{f'(a+\theta_1 h)}{\phi'(a+\theta_2 h)} = \frac{f'(a)}{\phi'(a)}, \quad \text{provided} \quad \phi'(a) \neq 0.$$

If, in addition to $f(a) = 0$ and $\phi(a) = 0$, we have $f'(a) = 0$ and $\phi'(a) = 0$, then

$$\lim_{x \to a} \frac{f(x)}{\phi(x)} = \lim_{h \to 0} \frac{f(a+h)}{\phi(a+h)}$$

$$= \lim_{h \to 0} \frac{f(a) + hf'(a) + \dfrac{h^2}{2!} f''(a+\theta_3 h)}{\phi(a) + h\phi'(a) + \dfrac{h^2}{2!} \phi''(a+\theta_4 h)}$$

$$= \lim_{h \to 0} \frac{f''(a+\theta_3 h)}{\phi''(a+\theta_4 h)} = \frac{f''(a)}{\phi''(a)},$$

provided $f''(x)$ and $\phi''(x)$ exist and are continuous in the neighbourhood of $x = a$, and $\phi''(a) \neq 0$.

Similarly, if in addition to the conditions already quoted $f''(a) = 0$ and $\phi''(a) = 0$, then

$$\lim_{x \to a} \frac{f(x)}{\phi(x)} = \frac{f^{(3)}(a)}{\phi^{(3)}(a)},$$

provided $f^{(3)}(x)$ and $\phi^{(3)}(x)$ exist and are continuous in the neighbourhood of $x = a$, and $\phi^{(3)}(a) \neq 0$. This process can be extended indefinitely. Thus to evaluate $\lim_{x \to a} \frac{f(x)}{\phi(x)}$, we evaluate $\frac{f'(a)}{\phi'(a)}$, $\frac{f''(a)}{\phi''(a)}$, etc. until we obtain an answer *not* of the form $\tfrac{0}{0}$.

Example. Find
$$\lim_{x \to 1} \frac{x-1}{\log x}.$$

If $f(x) = x-1$, and $\phi(x) = \log x$, $f'(1) = 1$, $\phi'(1) = 1$. Hence

$$\lim_{x \to 1} \frac{x-1}{\log x} = 1.$$

Example. Find
$$\lim_{x \to 0} \frac{x - \sin x}{x - \tan x}.$$

If $f(x) = x - \sin x$ and $\phi(x) = x - \tan x$, it is easily seen that

$$f'(0) = 0, \qquad \phi'(0) = 0$$
$$f''(0) = 0, \qquad \phi''(0) = 0$$
$$f^{(3)}(0) = 1, \qquad \phi^{(3)}(0) = -2.$$

Hence
$$\lim_{x \to 0} \frac{x - \sin x}{x - \tan x} = -\tfrac{1}{2}.$$

It is often easier to use the Maclaurin expansions of the function.

THE MEAN VALUE THEOREM AND ITS APPLICATIONS 75

Example.

$$\frac{x-\sin x}{x-\tan x} = \frac{x-\left(x-\dfrac{x^3}{3!}+\dfrac{x^5}{5!}-\cdots\right)}{x-\left(x+\dfrac{x^3}{3}+\dfrac{2x^5}{15}+\cdots\right)}$$

$$= \frac{\dfrac{x^3}{6}-\dfrac{x^5}{120}+\cdots}{-\dfrac{x^3}{3}-\dfrac{2x^5}{15}-\cdots}$$

$$= -\frac{1}{2}\frac{\left(1-\dfrac{x^2}{20}+\cdots\right)}{\left(1+\dfrac{2}{5}x^2+\cdots\right)}$$

whence it is clear that

$$\lim_{x\to 0}\frac{x-\sin x}{x-\tan x} = -\frac{1}{2}.$$

Estimating errors in numerical calculations

We have considered in Chapter 2 the problem of estimating errors when we use infinite series for numerical calculations. The Mean Value theorem is useful in such an estimation. The magnitude of the error in taking the sum of the first n terms of the Maclaurin expansion of $f(x)$ to represent $f(x)$ is less than

$$\left|\frac{x^n}{n!}f^{(n)}(\theta x)\right|.$$

Example.

$$\sin(a+h) = \sin a + h\cos a + \frac{h^2}{2!}(-\sin a) +$$

$$+\frac{h^3}{3!}(-\cos a) + \frac{h^4}{4!}\sin(a+\theta h).$$

Clearly the error in calculating the value of $\sin(\tfrac{1}{4}\pi+0\cdot 1)$ by taking the first four terms of the Taylor expansion is less than $(0\cdot 1)^4/4!$. Thus

$$\sin(\tfrac{1}{4}\pi+0\cdot 1) = \frac{1}{\sqrt{2}}\left(1+0\cdot 1-\frac{1}{2!}(0\cdot 1)^2-\frac{1}{3!}(0\cdot 1)^3\right)$$

$$= 0\cdot 774\ 164,$$

with an error less than $0\cdot 000\ 005$.

Exercises 3(c)

1. Find the values of the following limits:

$$\lim_{x \to 0} \frac{\log(1+2x) - 2x + 2x^2}{x^2 \tan^{-1} x} ; \qquad \lim_{x \to 0} \frac{\log(1 + x^2 + x^4)}{x(e^x - 1)} ;$$

$$\lim_{x \to 1} \frac{2e^x - e^{x^2} - e}{(\log x)^2} ; \qquad \lim_{x \to \frac{1}{2}\pi} \frac{\cos^3 x}{(x - \frac{1}{2}\pi)^2 \cos 5x} .$$

2. Evaluate the following limits in two ways (i) using L'Hôpital's rules (ii) by using expansions in power series:

$$\lim_{x \to 0} \frac{x^2 + 2 \cos x - 2}{x^4} ; \qquad \lim_{x \to 1} \frac{\log x}{\cos(\frac{1}{2}\pi x)} ;$$

$$\lim_{x \to 0} \frac{\sin^{-1} x}{\sin(2x)} ; \qquad \lim_{x \to \frac{1}{2}} \frac{1 + \cos 2\pi x}{(2x - 1)^2} .$$

3. Find an upper bound for the error in calculating log(1·2) using the first five terms of the logarithm series.

4. Find an upper bound for the error in calculating $e^{\frac{1}{2}}$ using the first four terms of the exponential series.

Further applications of Taylor's theorem

Curvature

Let P be any point on a plane curve (see Fig. 3.8). Taking P as the origin of coordinates, the tangent at P as the x-axis and the inward drawn normal at P as the y-axis, let the equation of the curve be $y = f(x)$.

The curvature at P is $\dfrac{f''(0)}{[1 + \{f'(0)\}^2]^{\frac{3}{2}}}$, i.e. $f''(0)$, since $f'(0) = 0$.

The Maclaurin expansion of $f(x)$ in the neighbourhood of P is

$$f(x) = f(0) + xf'(0) + \frac{x^2}{2}f''(0) + \cdots,$$

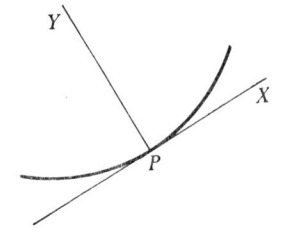

Fig. 3.8

i.e. since $f(0) = 0, f'(0) = 0$,

$$f(x) = \frac{x^2}{2}f''(0)+\cdots,$$

whence

$$f''(0) = \lim_{x \to 0} \frac{2f(x)}{x^2}.$$

Hence the curvature at P is

$$\lim_{x \to 0} \frac{2f(x)}{x^2} = \lim_{x \to 0} \frac{2y}{x^2},$$

a result due to Newton.

Example. The curvature of the parabola $x^2 = 4ay$ at 0 is

$$\lim_{x \to 0} \frac{2y}{x^2} = \frac{1}{2a}.$$

Contact of curves

Consider the intersections of the two curves $y = f(x)$, $y = \phi(x)$. The values of x at points of intersection satisfy the equation

$$f(x) - \phi(x) = 0.$$

Let us denote $f(x) - \phi(x)$ by $F(x)$, and let $x = a$ be any root of the equation $F(x) = 0$ and P the corresponding point of intersection of the two curves.

Then the equation $F(x) = 0$ may have a *single* root $x = a$, in which case the two curves cut at an angle at P (Fig. 3.9). Or the equation $F(x) = 0$ may have a double root at $x = a$, in which case $F'(a) = 0$ in addition to $F(a) = 0$. It follows that $f'(a) = \phi'(a)$, which means that the two curves will touch at P (Fig. 3.10).

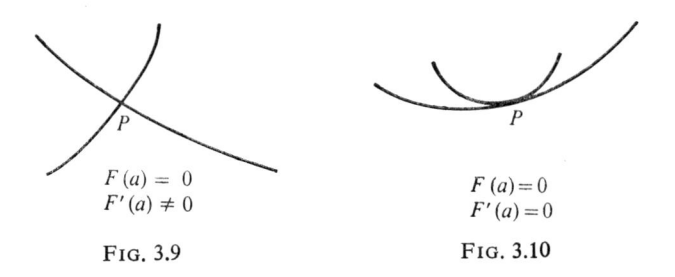

$F(a) = 0$
$F'(a) \neq 0$

FIG. 3.9

$F(a) = 0$
$F'(a) = 0$

FIG. 3.10

Again, the equation $F(x) = 0$ may have a triple root $x = a$, in which case $F(a) = 0$, $F'(a) = 0$, $F''(a) = 0$, whence $f(a) = \phi(a)$, $f'(a) = \phi'(a)$, $f''(a) = \phi''(a)$. We say that the curves have a three-point contact at P; we cannot distinguish graphically between a two- and a three-point contact, it is a purely algebraic concept.

In general, the two curves $y = f(x)$ and $y = \phi(x)$ are said to have an r-point contact at $x = a$ if

$$f(a) = \phi(a), f'(a) = \phi'(a), \ldots f^{(r-1)}(a) = \phi^{(r-1)}(a).$$

Example. Find the equation of the circle which has closest contact with the parabola $y^2 = 4x$ at the point $P(1, 2)$.

Consider the contact of the parabola with the circle

$$x^2 + y^2 + 2gx + 2fy + c = 0.$$

There is a one-point contact at P if

$$1 + 4 + 2g + 4f + c = 0. \tag{1}$$

At P, $\dfrac{dy}{dx}$ for the parabola is 1, while for the circle

$$2x + 2y\frac{dy}{dx} + 2g + 2f\frac{dy}{dx} = 0.$$

Hence there is a two-point contact at P if

$$2 + 4(1) + 2g + 2f(1) = 0. \tag{2}$$

At P, $\dfrac{d^2y}{dx^2}$ for the parabola is $-\tfrac{1}{2}$, while for the circle

$$2 + 2y\frac{d^2y}{dx^2} + 2\left(\frac{dy}{dx}\right)^2 + 2f\frac{d^2y}{dx^2} = 0$$

Hence there is a three-point contact at P if

$$2 + 4(-\tfrac{1}{2}) + 2 + 2f(-\tfrac{1}{2}) = 0. \tag{3}$$

Thus the circle and the parabola will have a three-point contact at P if equations (1), (2), and (3) are satisfied. These equations have for solution $f = 2$, $g = -5$, $c = -3$. Thus the circle

$$x^2 + y^2 - 10x + 4y - 3 = 0$$

has a three-point contact with the parabola at P, and is the circle of closest contact with the parabola at P.

Find the radius of curvature of the parabola at P and show that it is equal to the radius of the circle of closest contact.

Exercises 3(d)

1. Find the values of a and b in order that the curve $ax^2 + by^2 = 12$ should have a two-point contact with the curve $y = 2x^4$ at the point $(1, 2)$.

2. Find the values of a, b, h in order that the conic

$$ax^2 + 2hxy + by^2 = 1$$

should have a three-point contact with the curve $x^3 + y^3 = 2xy$ at the point $(1, 1)$.

3. Find the radius of curvature of the ellipse $\dfrac{x^2}{a^2} + \dfrac{y^2}{b^2} = 1$ at the points $(a, 0)$, $(0, b)$ using Newton's rule.

4. Prove, using Newton's method, that the radius of curvature of the curve $x^2 y = a^2(a+y)$ at the point $(0, -a)$ is $\tfrac{1}{2}a$.

5. Prove that the circle of closest contact with a curve at any point P on the curve is the circle of curvature at P.

6. The parametric equations of a curve in space are given by

$$x = t^3, \quad y = 2t^2 + t, \quad z = t^4 + 1.$$

Find the equation in t giving the parameters of the points of intersection of a plane

$$ax + by + cz + d = 0$$

with the curve and find the equation of the plane which has closest contact with the curve at the point $(1, 3, 2)$ of the curve.

7. If $V = 4 \cos \theta - \cos 2\theta$, prove that V has a maximum value when $\theta = 0$.

8. Investigate the maximum and minimum values of

$$y = \cos x + \cosh x.$$

Remainder term in Taylor's series

We shall now give an alternative proof of the two forms of the remainder term in the Taylor series. It is clear that

$$f(x) = f(a) + \int_a^x f'(t)\, dt,$$

which gives, on integrating by parts,

$$f(x) = f(a) - [f'(t)(x-t)]_a^x + \int_a^x f''(t)(x-t)\,dt$$

$$= f(a) + (x-a)f'(a) + \int_a^x f''(t)(x-t)\,dt$$

$$= f(a) + (x-a)f'(a) - \left[f''(t)\frac{(x-t)^2}{2}\right]_a^x +$$

$$+ \int_a^x f^{(3)}(t)\frac{(x-t)^2}{2}\,dt$$

$$= f(a) + (x-a)f'(a) + \frac{(x-a)^2}{2}f''(a) + \int_a^x f^{(3)}(t)\frac{(x-t)^2}{2}\,dt.$$

This process can be carried on indefinitely and gives

$$f(x) = f(a) + (x-a)f'(a) + \cdots + \frac{(x-a)^{n-1}}{(n-1)!}f^{(n-1)}(a) +$$

$$+ \int_a^x f^{(n)}(t)\frac{(x-t)^{n-1}}{(n-1)!}\,dt.$$

The problem then becomes one of estimating the value of the integral. We have proved in an earlier chapter that $\int_a^b f(x)\phi(x)\,dx$ lies between $m \int_a^b \phi(x)\,dx$ and $M \int_a^b \phi(x)\,dx$, where m and M are the least and greatest values of $f(x)$ in the range (a, b), provided $\phi(x)$ is positive throughout the range (a, b). Now it is intuitively true that if $f(x)$ is continuous, any number lying between m and M can be expressed as $f(\xi)$, where ξ is a number lying between a and b.

Thus

$$\int_a^b f(x)\phi(x)\,dx = f(\xi)\int_a^b \phi(x)\,dx.$$

This result is also true if $\phi(x)$ is negative throughout the range. Applying this result to the integral

$$\int_a^x f^{(n)}(t)\frac{(x-t)^{n-1}}{(n-1)!}\,dt,$$

THE MEAN VALUE THEOREM AND ITS APPLICATIONS

and assuming that $f^{(n)}(t)$ is continuous in the range (a, x) we can express the integral as

$$f^{(n)}(\xi) \int_a^x \frac{(x-t)^{n-1}}{(n-1)!} \, dt = f^{(n)}(\xi) \frac{(x-a)^n}{n!},$$

where $a < \xi < x$, which is the Lagrange form of the remainder. Alternatively, we can express the integral as

$$f^{(n)}(\xi) \frac{(x-\xi)^{n-1}}{(n-1)!} \int_a^x dt = f^{(n)}(\xi) \frac{(x-\xi)^{n-1}}{(n-1)!} (x-a),$$

which, on writing $a = 0$, $\xi = \theta x$, gives

$$\frac{x^n(1-\theta)^{n-1}}{(n-1)!} f^{(n)}(\theta x),$$

the Cauchy form of the remainder in Maclaurin's theorem.

The error in Simpson's rule for the approximate value of an integral

If $f(x)$ has a continuous fourth derivative $f^{(4)}(x)$ satisfying

$$|f^{(4)}(x)| \leq M$$

in the interval $c-h \leq x \leq c+h$, then it follows from the mean value theorem that

$$f(c-h) = f(c) - hf'(c) + \frac{h^2}{2!} f''(c) - \frac{h^3}{3!} f^{(3)}(c) + \frac{h^4}{4!} f^{(4)}(\xi),$$

where $c-h < \xi < c$, and

$$f(c+h) = f(c) + hf'(c) + \frac{h^2}{2!} f''(c) + \frac{h^3}{3!} f^{(3)}(c) + \frac{h^4}{4!} f^{(4)}(\xi_1),$$

where $c < \xi_1 < c+h$. It follows that

$$\frac{h}{3}\{f(c-h) + 4f(c) + f(c+h)\} = 2hf(c) + \tfrac{1}{3}h^3 f^{(3)}(c) + R_1,$$

where

$$|R_1| \leq 2\frac{Mh^4}{4!} \cdot \frac{h}{3} = \frac{Mh^5}{36}.$$

Now
$$\int_{c-h}^{c+h} f(x)\,dx = \int_{c-h}^{c+h} \{f(c)+(x-c)f'(c)+\frac{(x-c)^2}{2!}f''(c)+$$
$$+\frac{(x-c)^3}{3!}f^{(3)}(c)+\frac{(x-c)^4}{4!}f^{(4)}(\xi_2)\}\,dx,$$

where $c-h < \xi_2 < c+h$,

$$= 2hf(c)+\tfrac{1}{3}h^3 f^{(3)}(c)+R_2,$$

where $|R_2| \leq M\int_{c-h}^{c+h}\frac{(x-c)^4}{4!}\,dx = M\frac{h^5}{60}$. Hence

$$\int_{c-h}^{c+h} f(x)\,dx = \frac{h}{3}\{f(c-h)+4f(c)+f(c+h)\}+R_3,$$

where $|R_3| \leq \frac{Mh^5}{36}+\frac{Mh^5}{60} = \frac{2Mh^5}{45}$. It follows that if $|f^{(4)}(x)| \leq M$ in the interval $a \leq x \leq b$, by dividing the interval into $2n$ equal parts of width h, an approximate value of $\int_a^b f(x)\,dx$ is

$$\frac{h}{3}\{f(a)+4f(a+h)+2f(a+2h)+\cdots+f(b)\},$$

with an error R, where $|R| \leq n\frac{2Mh^5}{45} = \frac{M(b-a)^5}{720n^4}$.

Example. $\log 2 = \int_2^4 x^{-1}\,dx$. By taking $n = 4$, $h = \tfrac{1}{4}$, we obtain an approximation to $\log 2$ in the form

$$\frac{\tfrac{1}{4}}{3}\left\{\frac{1}{2}+4\frac{1}{2+\tfrac{1}{4}}+2\frac{1}{2+\tfrac{1}{2}}+\cdots+\frac{1}{4}\right\} = 0\cdot693\,12.$$

Now $f^{(4)}(x) = \frac{4!}{x^5}$ and $|f^{(4)}(x)| \leq \frac{4!}{2^5} = \frac{3}{4}$, whence $|R| \leq 0\cdot000\,14$. It follows that $\log 2$ lies between $0\cdot693\,12 - 0\cdot000\,14$ and $0\cdot693\,13 + 0\cdot000\,14$, i.e. between $0\cdot692\,98$ and $0\cdot693\,27$.

THE MEAN VALUE THEOREM AND ITS APPLICATIONS

Exercises 3(e)

1. $\log(\tfrac{3}{2})$ is to be calculated by Simpson's rule applied to $\int_{2}^{3} x^{-1}\,dx$. What is the least number of intervals that will ensure an error not exceeding 10^{-5}? Hence calculate limits between which $\log(\tfrac{3}{2})$ must lie.

2. If $f(x)$ has a continuous second derivative $f''(x)$ satisfying $|f''(x)| \leq M$ in the interval $c-h \leq x \leq c+h$, prove that
$$\int_{c-h}^{c+h} f(x)\,dx = h\{f(c-h)+f(c+h)\} + R,$$
where $|R| \leq \dfrac{4Mh^3}{3}$.

If $|f''(x)| \leq M$ in the interval $a \leq x \leq b$, prove by dividing the interval into n equal parts of width $2h$, that an approximate value of $\int_a^b f(x)\,dx$ is
$$h\{f(a)+2f(a+2h)+\cdots +f(b)\}$$
with an error R, where $|R| < n\dfrac{4Mh^3}{3} = \dfrac{M(b-a)^3}{6n^2}$.

Use this result to estimate the value of $\log 2$ using $\int_1^2 x^{-1}\,dx$ and taking n large enough to ensure that the error is less than $0\cdot 01$. Hence show that $0\cdot 685 < \log 2 < 0\cdot 705$.

3. By approximating the graph of the function $f(x)$ in the interval $[-h, h]$ by its tangent at $(0, f(0))$, show that $2hf(0)$ is an approximate value of $\int_{-h}^{h} f(x)\,dx$ with an error not greater than $\dfrac{Mh^3}{3}$, provided that $|f''(x)| \leq M$ in $[-h, h]$. Hence derive the approximation $2h\sum_{r=1}^{n} f_r$ for $\int_a^b f(x)\,dx$, where $f_r = f(a+(2r-1)h)$, $h = \dfrac{b-a}{2n}$, and show that the error does not exceed $\dfrac{M(b-a)h^2}{6}$, provided that $|f''(x)| \leq M$ in $[a, b]$.

Use this result to estimate $\log 3$, taking h sufficiently small to ensure that the error is less than $0\cdot 01$, and so show that $1\cdot 088 < \log 3 < 1\cdot 106$.

Miscellaneous problems on Chapter 3

1. The functions $g(x)$ and $h(x)$ are continuous for $a \leq x \leq b$ and differentiable for $a < x < b$; k is a constant. Apply Rolle's theorem to the function
$$f(x) = g(x) - g(a) + k\{h(x) - h(a)\},$$
and prove that, provided $h'(x) \neq 0$ for $a < x < b$, there exists a number ξ, $a < \xi < b$, such that
$$\frac{g(b)-g(a)}{h(b)-h(a)} = \frac{g'(\xi)}{h'(\xi)}.$$

Hence establish L'Hôpital's rules for limits.

2. Find the values of the following limits in two ways, (i) by using the standard series for the functions concerned, (ii) by using L'Hôpital's rules.

$$\lim_{x \to 0} \frac{\sin x \sinh x}{\log^2(1+x)} \ ; \quad \lim_{x \to 0} \frac{x \sin x}{\cos x - \cosh x} \ ;$$

$$\lim_{x \to \pi} \frac{(x-\pi)\sin x}{1+\cos x}.$$

3. Prove that if x is small and positive, $(1+x)^{-\frac{1}{2}}$ is approximately equal to $1 - \frac{1}{2}x$, with an error which is less than $\frac{3}{8}x^2$. By considering the integral
$$\int_0^x \frac{dy}{\sqrt{(1+y^2)}},$$ prove that when $0 < x < 1$, $\log(x+\sqrt{(x^2+1)})$ lies between $x - \frac{x^3}{6}$ and $x - \frac{x^3}{6} + \frac{3}{40}x^5$.

4. Prove that if $f(x) = e^x \cos x$, $f^{(n)}(x) = 2^{\frac{1}{2}n}e^x \cos(x + \frac{1}{4}n\pi)$. Justify the existence of the Maclaurin expansion of $e^x \cos x$ for all values of x and find the coefficient of x^n in the expansion.

5. Prove, using Newton's rule, that the radius of curvature of the curve
$$a^2 y^2 = x^3(a-x)$$
at the point $(a, 0)$ is $\frac{1}{2}a$.

6. Prove that the Maclaurin expansion of $\log(1+e^x)$ as far as the term in x^4 is
$$\log(1+e^x) = \log 2 + \tfrac{1}{2}x + \tfrac{1}{8}x^2 - \tfrac{1}{192}x^4 \ldots.$$

Prove that the Maclaurin expansion of $\tan^{-1}(1+x)$ as far as the term in x^3 is
$$\tan^{-1}(1+x) = \frac{1}{4} + \frac{x}{2} - \frac{x^2}{4} + \frac{x^3}{12} \ldots.$$

7. Calculate the following limits:
$$\lim_{t \to 0} (\cos t)^{1/t^2}; \quad \lim_{y \to 0} \frac{\sin 2y - 2y}{\tan^{-1} y - y}.$$

8. Prove that $\log(1+x) = x - \frac{x^2}{2(1+\theta x)^2}$, where θ lies between 0 and 1. Hence, or otherwise, prove that if $x > 0$
$$\lim_{n \to \infty} \frac{n!}{(x+1)(x+2) \cdots (x+n)} = 0.$$

9. Prove that if $f(x)$ and $\phi(x)$ are continuous in the range $[a, b]$ and differentiable in the range (a, b) then a number ξ exists in the range (a, b) such that
$$\begin{vmatrix} f(a) & f(b) \\ \phi(a) & \phi(b) \end{vmatrix} = (b-a) \begin{vmatrix} f(a) & f'(\xi) \\ \phi(a) & \phi'(\xi) \end{vmatrix}.$$

10. Prove that, if $f(x)$ is a polynomial of degree at most 3, then
$$\int_0^{2h} f(x)\, dx = \tfrac{1}{3}h\{f(0)+4f(h)+f(2h)\}.$$

If $f(x)$ is any function with a bounded derivative of the fourth order, deduce that
$$\int_0^{2h} f(x)\, dx = \tfrac{1}{3}h\{f(0)+4f(h)+f(2h)\}+R,$$

where $|R| < k\,|h|^5$, the constant k being independent of h.

Apply the approximation
$$\int_0^{2h} f(x)\, dx = \tfrac{1}{3}h\{f(0)+4f(h)+f(2h)\}$$

to the integral $\int_0^{\frac{1}{2}} \dfrac{dx}{1+x^2}$ and find the error in using this approximation.

11. Prove that ξ exists such that
$$\cos a - \cos b = \tan \xi (\sin a - \sin b),$$

where $a < \xi < b$.

12. A well-known approximation for $I = \int_{-h}^{h} f(x)\, dx$ is given by Simpson's rule,
$$I = \frac{h}{3}\{f(-h)+4f(0)+f(h)\}.$$

The result is exact if $f(x)$ is a polynomial of degree 3.

By expressing $f(x)$ in the form $f(x) = \phi(x)+R(x)$, where $\phi(x)$ is a cubic polynomial and $R(x)$ is Lagrange's form of the remainder, prove that, if $|f^{(4)}(x)| \leq M$ in the range $(-h, h)$, the modulus of the error in Simpson's approximation for I is less than $\dfrac{2Mh^5}{45}$.

Hence show that if $\int_a^b f(x)\, dx$ is evaluated by means of Simpson's rule with the range (a, b) divided into $2n$ equal parts, the modulus of the error does not exceed $\dfrac{M(b-a)^5}{720n^4}$.†

† By a more refined method of estimating the error, the result for the maximum error can be improved to $\dfrac{M(b-a)^5}{2880n^4}$.
(See Hardy, G. H. (1959) *Course of Pure Mathematics*, p. 307, question 46. C.U.P.)

13. By dividing the range of integration into eight parts and applying Simpson's rule, find an approximation for $\int_0^4 \frac{1}{(1+x)}\,dx$. Use the result of question 12 to show that log 5 lies between $\frac{6089}{3780} \pm \frac{2}{15}$.

14. Evaluate the following limits:

$$\lim_{x \to 0} \frac{x - \sin x}{2x - \tan 2x}\,; \quad \lim_{t \to 0} \frac{(1+\frac{1}{2}t)^{\frac{1}{2}} - (1+\frac{1}{3}t)^{\frac{1}{2}}}{(1-\frac{1}{2}t)^{-\frac{1}{2}} - (1-\frac{1}{3}t)^{-\frac{1}{2}}}.$$

15. By dividing the range of integration into four equal parts and applying Simpson's rule find an approximate value of

$$\int_0^1 \frac{dx}{1+x^2}.$$

Prove that the modulus of the fourth differential coefficient of $1/(1+x^2)$ is less than 24 in the range $(0, 1)$, and hence estimate the error in the approximation using the result of question 12.

16. Prove that for small values of h, an approximate value of $\tan(\frac{1}{4}\pi + h)$ is $1 + 2h + 2h^2 + \frac{8}{3}h^3$.

17. If $f(x)$, $\phi(x)$ and $\lambda(x)$ are functions of x continuous and differentiable in the closed interval $[a, b]$, prove that there is a number ξ in the range (a, b) such that

$$\begin{vmatrix} f(a) & \phi(a) & \lambda(a) \\ f(b) & \phi(b) & \lambda(b) \\ f'(\xi) & \phi'(\xi) & \lambda'(\xi) \end{vmatrix} = 0.$$

18. If $f(x) = \frac{x^2}{1+x}$, show that in the range $[1, 2]$ $|f^{(4)}(x)| < \frac{3}{4}$. By using Simpson's rule to approximate to the integral $\int_1^2 \frac{x^2}{1+x}\,dx$, show that $0\cdot 404 < \log(\frac{3}{2}) < 0\cdot 407$.

19. Use Simpson's rule to find an approximate value of the integral

$$I = \int_2^3 \frac{x^4}{1+x}\,dx, \text{ taking } h = \tfrac{1}{4}.$$ Prove that the modulus of the error is

less than $0\cdot 000\,085$ and show that $11\cdot 7042 < I < 11\cdot 7045$. Evaluate the integral and hence show that $0\cdot 2875 < \log(\frac{4}{3}) < 0\cdot 2879$.

4
Techniques of Integration

Techniques of integration

The reader will be already aware of the value of reduction formulae in the technique of integration. In simple cases, e.g. $I_n = \int_0^{\frac{1}{2}\pi} \sin^n \theta \, d\theta$, we obtain a relation between I_n and I_{n-2}, namely, $I_n = \frac{n-1}{n} I_{n-2}$, which enables us to find I_n in terms of I_1 or I_0. In more complicated cases we may find a relation between I_n, I_{n-1}, and I_{n-2}.

Example. If
$$I_n = \int_0^\pi \frac{\cos nx}{5 - 4\cos x} \, dx,$$
then
$$2I_n - 5I_{n-1} + 2I_{n-2} = \int_0^\pi \frac{2(\cos nx + \cos \overline{n-2}\,x) - 5\cos \overline{n-1}\,x}{5 - 4\cos x} \, dx$$

$$= \int_0^\pi \frac{4\cos x \cos \overline{n-1}\,x - 5\cos \overline{n-1}\,x}{5 - 4\cos x} \, dx$$

$$= -\int_0^\pi \cos \overline{n-1}\,x \, dx = -\left[\frac{\sin \overline{n-1}\,x}{n-1}\right]_0^\pi = 0,$$

provided $n \neq 1$. Thus $2I_n - 5I_{n-1} + 2I_{n-2} = 0$. A relation of this kind is called a difference equation and can be treated in the following way.

Consider the infinite series $S = I_0 + I_1 t + I_2 t^2 + \cdots$, where t is chosen so that the series converges. It could be argued that there is no guarantee that such a value of t exists, but the procedure may be regarded as one which *suggests* a result for I_n, which may then be proved formally by induction.

$$2S = 2I_0 + 2I_1 t + 2I_2 t^2 + \cdots,$$
$$-5tS = -5I_0 t - 5I_1 t^2 - \cdots,$$
$$2t^2 S = 2I_0 t^2 + \cdots.$$

Hence
$$(2-5t+2t^2)S = 2I_0 + (2I_1 - 5I_0)t.$$

We find that $I_0 = \pi/3$ and $I_1 = \pi/6$ by elementary methods. Thus

$$(2-5t+2t^2)S = \frac{2\pi}{3}(1-2t),$$

and hence

$$S = \frac{2\pi}{3} \cdot \frac{1-2t}{(2-t)(1-2t)} = \frac{\pi}{3}(1-\tfrac{1}{2}t)^{-1} = \frac{\pi}{3}\sum_0^\infty (\tfrac{1}{2}t)^n.$$

It follows that $I_n = \frac{\pi}{3}(\tfrac{1}{2})^n$, provided $n \neq 1$. But the result proved for I_1 show that the expression for I_n is also valid for $n = 1$. Thus, the result $I_n = \frac{\pi}{3}(\tfrac{1}{2})^n$ is valid for all positive integral values of n.

Exercises 4(a)—Difference equations

1. If $v_1 = 1$, $v_2 = 2$, and $v_n - 2v_{n-1} - 15v_{n-2} = 0$ for $n \geq 3$, prove that $v_n = \tfrac{1}{8}\{5^n - (-3)^n\}$.

2. From 1 seed there grows a plant which produces 3 seeds at the beginning of the second year, 18 seeds at the beginning of the third year, and dies at the end of the third year. Calculate how many seeds will be produced at the beginning of the nth year.

3. The sequence $1, 3, 2, \tfrac{5}{2}, \ldots$ is defined by the fact that from the third term onwards, each term is the arithmetic mean of the two preceding terms. Show that the limiting value of the nth term of the series as n tends to infinity is $\tfrac{7}{3}$.

4. A set of terms $u_1, u_2, \ldots$ is formed according to the following rules: $u_1 = 1$, $u_2 = 3$, $u_3 = -4$, $u_{n+3} = 3u_{n+1} - 2u_n$. Find u_n in terms of n.

5. In the sequence of terms $1, 1, 2, 3, 5, 8, 13, \ldots$, each term is the sum of the previous two terms. Prove that the nth term of the sequence is
$$\frac{1}{\sqrt{5}}\left\{\left(\frac{1+\sqrt{5}}{2}\right)^n - \left(\frac{1-\sqrt{5}}{2}\right)^n\right\},$$
and that the limiting value as n tends to infinity of the ratio of the nth term to the $(n-1)$th term is $\tfrac{1}{2}(\sqrt{5}+1)$.

6. Starting with a single cutting of a plant, a gardener takes one cutting after it has been growing 1 year, and 9 cuttings each year thereafter; he treats the cuttings in the same way. Prove that in the nth year of growth of the original plant, the total number of plants and cuttings is

$$\tfrac{2}{3}(4)^{n-1}+\tfrac{1}{3}(-2)^{n-1}.$$

7. If $\dfrac{u_n}{v_n}$ is the fraction obtained by evaluating the continued fraction

$$\dfrac{2}{3+}\dfrac{2}{3+}\dfrac{2}{3+}\cdots \text{ as far as } n \text{ terms, find difference formulae for } u_n \text{ and } v_n.$$

Hence show that, as n tends to infinity, u_n/v_n tends to $\tfrac{1}{2}(\sqrt{17}-3)$.

8. Solve the equation
$$u_{n+2}+2u_{n+1}+u_n = 5.2^n,$$
where $u_0 = 2$, $u_1 = 1$.

Exercises 4(b)—Reduction formulae for integrals

1. Find a reduction formula for $I_n = \int x^n e^x \, dx$. Hence evaluate $\int_0^1 x^4 e^x \, dx$.

2. Find a reduction formula for $I_n = \int x^n \cos x \, dx$ and evaluate I_2 and I_3.

3. Evaluate the following integrals:

$$\int_0^2 x^3 e^{4x} \, dx; \quad \int_0^{\frac{1}{2}\pi} x^4 \sin x \, dx; \quad \int_0^{\frac{1}{2}\pi} x^3 \sin x \, dx.$$

4. If $I_n = \int \dfrac{\sin n\theta}{\sin \theta} \, d\theta$, prove that

$$I_n - I_{n-2} = \dfrac{2 \sin(n-1)\theta}{n-1},$$

and evaluate $\displaystyle\int_0^{\frac{\pi}{4}} \dfrac{\sin 5\theta}{\sin \theta} \, d\theta$ and $\displaystyle\int_0^{\frac{\pi}{2}} \dfrac{\sin 6\theta}{\sin \theta} \, d\theta$.

5. If $I_n = \int x^n(1+x^4)^{-\frac{1}{2}} \, dx$, prove, by applying the method of integration by parts to the integral $I_n + I_{n+4}$ that

$$I_n + \dfrac{n+3}{n+1} I_{n+4} = \dfrac{x^{n+1}}{n+1} (1+x^4)^{\frac{1}{2}}.$$

Evaluate the integral $\int_0^1 x^n(1+x^4)^{-\frac{1}{2}} \, dx$ in the cases $n = 1$, $n = 5$; $n = 3$, $n = 7$.

6. Find a reduction formula for $I_n = \int \sec^n x \, dx$ and evaluate the integrals

$$\int_0^{\frac{\pi}{4}} \sec^5 x \, dx; \int_0^{\frac{\pi}{3}} \sec^8 x \, dx.$$

7. If $I_n = \int \dfrac{\sin(2n-1)\theta}{\sin\theta}\, d\theta$ find the value of $I_n - I_{n-1}$. Hence show that, when n is a positive integer,

$$\int_0^{\frac{1}{2}\pi} \dfrac{\sin(2n-1)\theta}{\sin\theta}\, d\theta = \tfrac{1}{2}\pi \quad \text{and} \quad \int_0^{\frac{1}{2}\pi} \dfrac{\sin^2 n\theta}{\sin^2\theta}\, d\theta = \tfrac{1}{2}n\pi.$$

8. If $I_{m,n} = \int x^m (\log x)^n\, dx$, prove that

$$(m+1)I_{m,n} = x^{m+1}(\log x)^n - nI_{m,n-1},$$

and hence evaluate $\int_1^e x^3(\log x)^2\, dx$.

Example. Find a reduction formula for $I_n = \int (ax^2+2bx+c)^{-n}\, dx$, where $ac - b^2 \neq 0$.

On writing $x + \dfrac{b}{a} = t$, $I_n = a^{-n}\int (t^2+k)^{-n}\, dt$, where $k = \dfrac{ac-b^2}{a^2}$. Then

$$I_n = a^{-n}t(t^2+k)^{-n} - a^{-n}\int t(-n)(t^2+k)^{-n-1} 2t\, dt$$

$$= a^{-n}t(t^2+k)^{-n} + 2na^{-n}\int (t^2+k-k)(t^2+k)^{-n-1}\, dt$$

$$= a^{-n}t(t^2+k)^{-n} + 2n(I_n - kaI_{n+1}).$$

Hence

$$(2n-1)I_n - 2n\left(\dfrac{ac-b^2}{a}\right)I_{n+1} + \left(x+\dfrac{b}{a}\right)(ax^2+2bx+c)^{-n} = 0.$$

Exercises 4(c)

1. Prove that, if $I_{m,n} = \int_0^1 x^m(1+x^2)^{-n}\, dx$, $m \geq 1$, then

$$I_{m,n} = \dfrac{m-1}{2(n-1)} I_{m-2,n-1} - \dfrac{1}{2^n(n-1)}.$$

Hence evaluate $\int_0^1 x^4(1+x^2)^{-3}\, dx$.

2. If $I_n = \int \dfrac{x^n\, dx}{\sqrt{(ax^2+2bx+c)}}$, find a formula connecting I_n, I_{n-1} and I_{n-2}, n being a positive integer.

Hence, or otherwise, evaluate $\int_0^1 \dfrac{x^3}{\sqrt{(x^2+2x+2)}}\, dx$.

TECHNIQUES OF INTEGRATION

3. By use of a reduction formula, or otherwise, prove that if a and $ac-b^2$ are positive, and n is a positive integer, then

$$\int_{-\infty}^{\infty} \frac{dx}{(ax^2+2bx+c)^{n+1}} = \frac{1.3.5\cdots(2n-1)}{2.4.6\cdots(2n)} \cdot \frac{\pi a^n}{(ac-b^2)^{n+\frac{1}{2}}}.$$

4. If $\int x^m(a+bx^n)^p \, dx = u_p$, prove that

$$(np+m+1)u_p = x^{m+1}(a+bx^n)^p + anp\, u_{p-1}.$$

5. If $u_n = \int_0^{\frac{1}{2}\pi} x \cos^n x \, dx$, where $n > 1$, show that

$$u_n = -\frac{1}{n^2} + \frac{n-1}{n} u_{n-2}.$$

Evaluate u_4 and u_5.

6. Show that

$$\int \frac{(x-p)^{2n+1}}{(ax^2+2bx+c)^{n+\frac{3}{2}}} dx = \frac{1}{(b^2-ac)^{n+1}} \int (y^2-q^2)^n \, dy,$$

where $q^2 = ap^2+2bp+c$, and $y(ax^2+2bx+c)^{\frac{1}{2}} = (ap+b)x+(bp+c)$.
Deduce that, when a, c, and $b+\sqrt{(ac)}$ are positive,

$$\int_0^\infty \frac{dx}{(ax^2+2bx+c)^{\frac{3}{2}}} = \frac{1}{\{b+\sqrt{(ac)}\}\sqrt{c}}$$

and

$$\int_0^\infty \frac{x \, dx}{(ax^2+2bx+c)^{\frac{3}{2}}} = \frac{1}{\{b+\sqrt{(ac)}\}\sqrt{a}}.$$

Simply- and multiply-connected regions

A simple closed curve in the xy-plane is a closed curve which does not intersect with itself at any point, e.g. circles, ellipses and astroids are simple closed curves; the curve $y^2 = x^2(1-x^2)$ is a closed curve but not a simple closed curve.

Simple closed curves are defined by the parametric equations $x = f(t)$, $y = \phi(t)$, where $f(t)$ and $\phi(t)$ are single-valued and continuous functions of t such that as t steadily increases or decreases from t_1 to t_2, the point $P(x, y)$ makes a single complete circuit of the curve.

In relation to any simple closed curve C, the set of points in the xy-plane can be divided into three categories:

(i) the set of points interior to C,
(ii) the set of points on the boundary C,
(iii) the set of points exterior to C.

The set of points interior to C is called an open region; the set of points interior to C or on the boundary C is called a closed region. Such regions can be represented analytically. For example, the

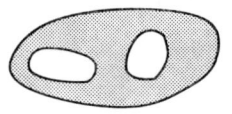

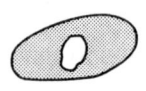

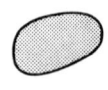

FIG. 4.1

open region interior to the circle centre O, radius a, is the set of points (x, y) satisfying $x^2+y^2 < a^2$. The closed region which also includes the boundary is the set of points (x, y) satisfying

$$x^2+y^2 \leq a^2.$$

More generally, a region is a set of points such that any two points of the set can be joined by a path which consists wholly of points of the set. A region may be bounded by a number of simple closed curves. The shaded areas shown in Fig. 4.1 are regions.

We categorize regions into simply-connected regions and multiply-connected regions. If any simple closed curve within a region can be shrunk to a point which also lies within the region, the region is called simply-connected; the region $x^2+y^2 < 1$ is a simply-connected region. But if we cannot shrink all curves of the region we call the region multiply-connected; the region consisting of the ring defined by $x^2+y^2 < 2$, $x^2+y^2 > 1$ is a multiply-connected region. We can shrink some curves of the region in Fig. 4.2, e.g. C_1, to points but not all curves of the region, e.g. C_2.

Descriptive properties of sets of points is the sphere of topology, a subject which has in recent times been developed in connection with the theory of sets.

Exercises 4(d)

1. Shade the regions in the xy-plane defined by the following inequalities and state whether the regions are simply-connected or multiply-connected.

(i) $x^2+4y^2 \leq 4$, $x+y \geq 1$.
(ii) $(x^2+4y^2-4)(x+y-1) \geq 0$.

TECHNIQUES OF INTEGRATION 93

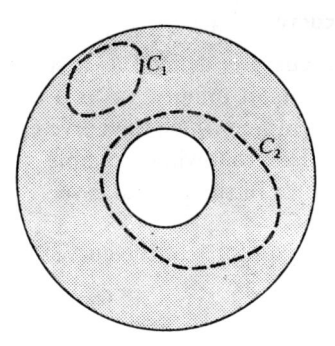

FIG. 4.2

2. Shade the region of the xy-plane in which the inequality
$$(x^2-3y^2-4)(xy-8) \geq 0 \qquad (a)$$
holds.
Indicate the points of the region for which
$$y^2+x = 8, \qquad (b)$$
and find the greatest value of x for points satisfying conditions (a) and (b) simultaneously.

3. Shade the region of the xy-plane in which the inequalities
$$\left.\begin{array}{r}x^2+y^2-4x-6y+9 \leq 0, \\ 3x+y \leq 11, \\ 3y-x \leq 9,\end{array}\right\} \qquad (a)$$
are satisfied simultaneously. By considering the family of lines $x+y = t$, where t is a real parameter, find the maximum value of $x+y$ subject to the conditions (a).
Find the minimum value of x^2+y^2 subject to the same conditions.

4. Shade the region in the xy-plane in which the inequality
$$(xy-1)(xy+2)(x+2y-2) \geq 0$$
is satisfied.
Find the radius of the largest circle, centre the origin, which lies completely within the region.

5. Shade the region in the xy-plane in which the inequality
$$(y^2-4x)(x^2+4y^2-4)(x-y) \leq 0$$
is satisfied.
Show that the distance of the point of the region nearest to the point $(\tfrac{5}{4}, 0)$ is $\sqrt{(69)}/12$.

Asymptotes of a curve

An asymptote of a curve is defined as a straight line which meets the curve in two points at infinity, or as a tangent to the curve at infinity, or as a straight line which the curve approaches as x and/or y tend to infinity. The following example shows the method of determining asymptotes in the general case.

Example. Find the asymptotes of the curve

$$x^3+2x^2y-xy^2-2y^3+4xy+2y^2+3x-17y-17 = 0, \qquad (1)$$

and sketch the curve.

The line
$$y = mx+c, \qquad (2)$$
meets the curve (1) where

$$x^3(1+2m-m^2-2m^3)+x^2(2c-2mc-6m^2c+4m+2m^2)+$$
$$+x(\cdots)+(\cdots) = 0.$$

The line (2) will be an asymptote to the curve (1) if this cubic equation in x has two infinite roots.

Now, on putting $x = 1/t$ in the equation $ax^3+bx^2+cx+d = 0$, the equation becomes $a+bt+ct^2+dt^3 = 0$, and the cubic equation in x will have two infinite roots if the corresponding cubic equation in t has two zero roots, i.e. if $a = 0$ and $b = 0$. Hence the line (2) will be an asymptote to the curve (1) if

$$1+2m-m^2-2m^3 = 0, \qquad (3)$$
and
$$2c-2mc-6m^2c+4m+2m^2 = 0. \qquad (4)$$

Equation (3) gives $m = 1, -1, -\frac{1}{2}$. From equation (4), when $m = 1$, $c = 1$; when $m = -1$, $c = -1$; when $m = -\frac{1}{2}$, $c = 1$. Thus we obtain three asymptotes, namely, $y = x+1$, $y = -x-1$, $y = -\frac{1}{2}x+1$. The combined equation of the three asymptotes can be written in the form

i.e.
$$(x-y+1)(x+y+1)(x+2y-2) = 0,$$
$$x^3+2x^2y-xy^2-2y^3+4xy+2y^2-3x+2y-2 = 0.$$

It will be noticed that the combined equation of the asymptotes can be written in a form which is precisely the same as the equation of the curve in respect of the terms of the two highest degrees.

TECHNIQUES OF INTEGRATION

Thus the equation of the curve can be written in the form

$$(x-y+1)(x+y+1)(x+2y-2) = -6x+19y+15,$$

say $L_1 L_2 L_3 = L_4$, a form which is most useful in sketching the curve.

The lines $L_1 = 0$, $L_2 = 0$, $L_3 = 0$, $L_4 = 0$ divide the xy-plane into regions. At the point O,

$$L_1 > 0, \quad L_2 > 0, \quad L_3 < 0, \quad L_4 > 0;$$

hence $L_1 L_2 L_3 < 0$ and $L_4 > 0$ throughout the region in which O lies. It follows that it is impossible for any portion of the curve to

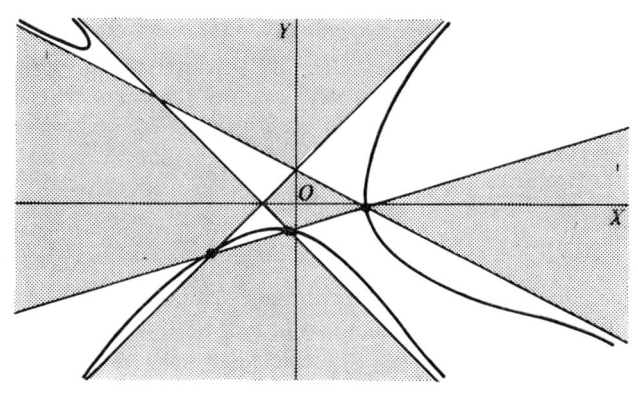

FIG 4.3

lie in this region. If by crossing a single boundary line of this first region, we enter a second region, then one of the factors L_1, L_2, L_3, or L_4 changes sign and the second region is a possible one for the curve to lie in. We can now shade regions of the plane in which it is impossible for the curve to lie.

Clearly, the points where the line $L_4 = 0$ cuts the lines $L_1 = 0$, $L_2 = 0$, $L_3 = 0$ are points on the curve. The curve may now be graphed (see Fig. 4.3).

The above example illustrates the truth of a general theorem that if an algebraic curve of degree n has n distinct asymptotes the combined equation of the asymptotes can be written in a form in which the terms of the two highest degrees i.e. the terms of degree n and $(n-1)$, are identical with the terms of the two highest degrees in the equation of the curve. Conversely, if the terms of the two highest degrees in the equation of an algebraic curve can be resolved into unrepeated linear factors $L_1, L_2, L_3...$, then the lines $L_1 = 0$, $L_2 = 0$, $L_3 = 0,...$ are asymptotes of the curve.

Notice that the method of finding asymptotes of a curve by considering the intersections of the curve with the line $y = mx+c$ will give all the asymptotes with the exception of those parallel to OY, i.e. asymptotes of the form $x = k$. A curve of degree n has normally n distinct asymptotes and if the method of using

$$y = mx+c$$

should yield fewer than n asymptotes, the possibility of asymptotes of the form $x = k$ should always be investigated.

Example. For the cubic curve $y^2(1-x) = x^2(2-x)$, we obtain two asymptotes by considering the intersection of the curve with the line $y = mx+c$, viz. $y = x-\frac{1}{2}$, $y = -x+\frac{1}{2}$.

The line $x = k$ meets the curve where $y^2(1-k) = k^2(2-k)$, which becomes, on writing $y = 1/t$,

$$k^2(2-k)t^3 - (1-k)t = 0.$$

For an asymptote, two roots of this *cubic* equation in t must be 0; one root is already 0, the second root will be 0 if $k = 1$. Thus we obtain a third asymptote $x = 1$, which is otherwise obvious if we write the equation of the curve in the form $y^2 = \dfrac{x^2(2-x)}{1-x}$; as $x \to 1, y \to \infty$.

Alternatively, we can investigate asymptotes of the form

$$x = my+c,$$

which will cover all the oblique asymptotes and those parallel to OY, thus giving a useful check of the oblique asymptotes.

Exercises 4(e)

1. Find the asymptotes of the curve whose equation is

$$x^3 - xy^2 + x^2 + y^2 + x + y = 0$$

2. Prove that no part of the curve

$$y^2(8-4x) = x^2(8-x)$$

lies between the lines $x = 2$ and $x = 8$, and that the line $x - 2y = 3$ is an asymptote of the curve.

Find the equations of the other asymptotes and of the tangents to the curve at the origin.
Sketch the curve.

3. Find the asymptotes of the curve
$$y^2 - x^2 - 3x - 5y + 7 = 0$$
and sketch the curve.

4. Find the asymptotes of the curve
$$x^3 + x^2y - 2xy^2 - 7x^2 + 6y^2 - 11xy + 8x + 15y + 3 = 0.$$
Denoting the three asymptotes by $L_1 = 0$, $L_2 = 0$, $L_3 = 0$, write the equation of the curve in the form $L_1 L_2 L_3 = L_4$. Sketch the curve.

5. Show that no part of the curve
$$y^2(x^2 - 4) = x^2(x^2 - 1)$$
lies between the lines $x = 1$ and $x = 2$.

Give the equations of (i) any axes of symmetry of the curve; (ii) the asymptotes; (iii) the tangents at the origin.
Sketch the curve.

6. Find the asymptotes of the curve
$$3x^2y + 2xy^2 - y^3 + 2y^2 + 6x^2 + 8xy + 7x + 6y + 4 = 0.$$
Sketch the curve.

7. Find the equations of the asymptotes of the curve
$$x^4 - y^4 + 16x^2 - 30y^2 - 225 = 0$$
and sketch the curve.

The area bounded by the curve and the line $x = \tfrac{15}{2}$ is rotated about OX; prove that the volume generated is $2293\pi/24$.

8. Sketch the curve $(x+y)(x-y)(x+y+1)^2 = 2x - y + 3$.

9. Find the equation of the real asymptote of the curve
$$x^3 + y^2x + x^2 - y^2 = 0,$$
and the equations of the tangents to the curve at 0. Sketch the curve.

10. Sketch the curve $(x-y+2)^2(2x-y+1)^2 = x+y$.

Equations of curves

A curve can be expressed analytically in various forms, including the Cartesian equation, the polar equation, or by a pair of parametric equations. In the remainder of this chapter we shall consider other analytical forms for a curve and deal with some miscellaneous problems in this field.

Polar equations and tangential polar equations

Length of a curve whose equation is given in polar coordinates

If $P(r, \theta)$ and $Q(r+\delta r, \theta+\delta\theta)$ are two neighbouring points on the curve (see Fig. 4.4),

whence
$$PQ = \sqrt{\{(r \sin \delta\theta)^2 + (r + \delta r - r \cos \delta\theta)^2\}},$$

and
$$\frac{PQ}{\delta\theta} = \sqrt{\left[r^2\left(\frac{\sin \delta\theta}{\delta\theta}\right)^2 + \left\{r\left(\frac{1 - \cos \delta\theta}{\delta\theta}\right) + \frac{\delta r}{\delta\theta}\right\}^2\right]}$$

but
$$\lim_{\delta\theta \to 0} \frac{PQ}{\delta\theta} = \sqrt{\left\{r^2 + \left(\frac{dr}{d\theta}\right)^2\right\}}$$

$$\lim_{\delta\theta \to 0} \frac{PQ}{\delta\theta} = \lim_{\delta\theta \to 0} \frac{PQ}{\delta s} \frac{\delta s}{\delta\theta} = \frac{ds}{d\theta}.$$

Hence
$$\frac{ds}{d\theta} = \sqrt{\left\{r^2 + \left(\frac{dr}{d\theta}\right)^2\right\}},$$

and
$$s = \int \sqrt{\left\{r^2 + \left(\frac{dr}{d\theta}\right)^2\right\}}\, d\theta.$$

It is also clear from the diagram that

$$\sin \angle PQN = \frac{r \sin \delta\theta}{PQ} = r \frac{\sin \delta\theta}{\delta\theta} \cdot \frac{\delta\theta}{\delta s} \cdot \frac{\delta s}{PQ},$$

and
$$\cos \angle PQN = \frac{r + \delta r - r \cos \delta\theta}{PQ}$$

$$= \left\{r\left(\frac{1 - \cos \delta\theta}{\delta\theta}\right) + \left(\frac{\delta r}{\delta\theta}\right)\right\} \frac{\delta\theta}{\delta s} \cdot \frac{\delta s}{PQ}.$$

If ϕ is the angle between the radius vector OP and the tangent to the curve at P, then as $Q \to P$, $\angle PQN \to \phi$. Hence $\sin \phi = r\dfrac{d\theta}{ds}$, $\cos \phi = \dfrac{dr}{ds}$ and $\tan \phi = r\dfrac{d\theta}{dr}$.

The tangential–polar equation of a curve

If P is any point on a curve (see Fig. 4.5), r the radius vector OP, and p the perpendicular from the origin O on the tangent at P, the

TECHNIQUES OF INTEGRATION 99

relation between p and r is called the tangential–polar equation (or the Pedal equation) of the curve.

An example of the way in which a p-r equation of a curve arises naturally is given by the problem of finding the path of a planet moving round the sun. If we take the sun as origin, the curve

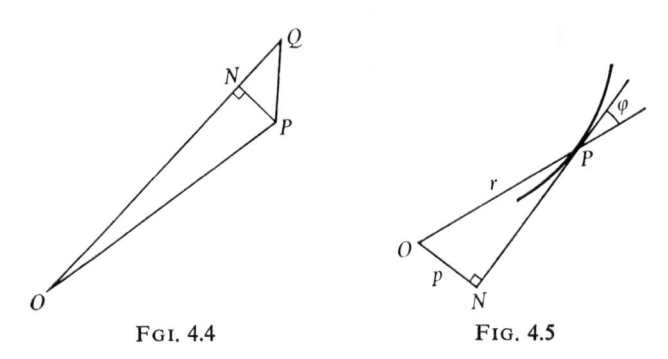

FIG. 4.4   FIG. 4.5

described by the planet is found by eliminating v, the velocity, from two dynamical equations, the energy equation and the moment of momentum equation, viz.

$$\tfrac{1}{2}mv^2 - \frac{m\mu}{r} = k, \quad (mv)p = mh,$$

where μ, h, k are constants, which gives the orbit in the form

$$\frac{h^2}{p^2} = \frac{2\mu}{r} + \frac{2k}{m}.$$

The p-r equation of a curve can easily be found from the polar equation of the curve. For let the polar equation of the curve be

$$f(r, \theta) = 0. \tag{1}$$

Then

$$p = r \sin \phi, \tag{2}$$

where

$$\tan \phi = r\frac{d\theta}{dr}. \tag{3}$$

We can eliminate θ and ϕ between equations (1), (2) and (3) to give the p-r equation of the curve.

Example. Consider the curve

$$r = a(2 - \cos \theta). \tag{1}$$

Then, on eliminating ϕ from

$$p = r \sin \phi, \quad \tan \phi = r\frac{d\theta}{dr} = \frac{r}{a \sin \theta},$$

we have

$$\frac{r^2}{p^2} = 1 + \frac{a^2 \sin^2 \theta}{r^2}.$$

Now, from eqn (1),

$$\sin^2 \theta = 1 - \left(2 - \frac{r}{a}\right)^2,$$

and hence the required p-r equation is

$$\frac{r^2}{p^2} = 1 + \frac{a^2}{r^2}\left\{1 - \left(2 - \frac{r}{a}\right)^2\right\},$$

i.e. $r^4 = p^2(4ar - 3a^2)$.

Exercises 4(f)

1. Sketch the spiral of Archimedes $r = a\theta$. Find an expression for the length of an arc of the spiral, measured from the origin. Prove that the pedal equation of the spiral is $r^4 = p^2(a^2 + r^2)$.

2. Find the p-r equation of the cardioid $r = a(1 + \cos \theta)$.

3. Sketch the curve $r(1 + \cos \theta) = a$, and find the length of an arc of the curve, measured from the point $(\tfrac{1}{2}a, 0)$.

4. If ϕ denotes the angle between the radius vector OP and the tangent to the curve at P, find the value of $\tan \phi$ for the curve $r^n = a^n \sin n\theta$. If a point P describes this curve in such a way that the angular velocity $d\theta/dt$ of the radius vector OP is a constant k, prove that the tangent at P rotates with angular velocity $(n+1)k$.

5. The polar equation of a conic referred to a focus as origin and the major axis as initial line is $l/r = 1 + e \cos \theta$, where l is the semi-latus rectum and e the eccentricity of the conic. Find the p-r equation of the conic.

6. Prove that, for the curve $r^n = a^n \cos n\theta$,

$$a^{2n}\frac{d^2r}{ds^2} + nr^{2n-1} = 0.$$

7. Sketch the curve $r \cos 5\theta = a$ and find its tangential-polar equation.

8. If $x = r \cos \theta$, $y = r \sin \theta$, where $f(r, \theta) = 0$, prove that

$$\left(\frac{dx}{d\theta}\right)^2 + \left(\frac{dy}{d\theta}\right)^2 = r^2 + \left(\frac{dr}{d\theta}\right)^2,$$

TECHNIQUES OF INTEGRATION

and hence establish the connection between the standard integrals for the length of an arc of a curve whose equation is given (i) in Cartesian form, (ii) in polar form.

9. Sketch the spiral $r = \dfrac{a}{\theta^2 - 1}$ ($\theta > 1$). Show that the length of the portion of the curve lying between the circles $r = a$, $r = b$ ($b > a$) is $\sqrt{\{b(a+b)\}} - a\sqrt{2}$.

A point P is taken on the curve and PQ is the bisector of the obtuse angle OPN between the radius vector OP and the outward drawn normal PN. The perpendicular to OP at O meets PQ at M. Show that OM is equal to the length of the arc OP of the curve.

10. If p is the length of the perpendicular from O on the tangent at any point (r, θ) on a curve whose polar equation is given, and $u = 1/r$, prove that

$$\frac{1}{p^2} = u^2 + \left(\frac{du}{d\theta}\right)^2.$$

11. Find the polar equations of the curves whose p-r equations are

  (i) $p = r^2$;   (ii) $p^3 = r^5$.

12. Sketch the curve $r = a \cos^3\left(\dfrac{\theta}{3}\right)$ and find the length of the curve.

13. Show that the total length of the curve $r = 2a \cos 3\theta$ is equal to the perimeter of the ellipse $x^2 + 9y^2 = 9a^2$.

14. Sketch the curve $p^2(4a^2 - 3r^2) = r^4$. Find the polar equation of the curve and the area of a loop of the curve.

15. Prove that any two straight lines drawn through the origin intercept arcs of the same length on the parabola $ay = x^2$ and the curve $r = a \sec^2 \theta$. Find this common length when the lines are $y = 0$, $y = x$.

16. Sketch the curve $r^2 = a^2 \cos 2\theta$ and find the $p-r$ equation of the curve.

If N is the foot of the perpendicular from O on the tangent at any point of a curve, the locus of N is called the pedal of the given curve. Show that the p-r equation of the pedal of the above curve is $a^2 p^3 = r^5$.

Radius of curvature in polar coordinates

In Fig. 4.6, it is clear that the inclination of the tangent to OX is

$$\psi = \theta + \phi = \theta + \cot^{-1}\left(\frac{1}{r}\frac{dr}{d\theta}\right).$$

Hence
$$\frac{d\psi}{d\theta} = 1 - \frac{1}{1 + \frac{1}{r^2}\left(\frac{dr}{d\theta}\right)^2}\left\{\frac{1}{r}\frac{d^2r}{d\theta^2} - \frac{1}{r^2}\left(\frac{dr}{d\theta}\right)^2\right\}.$$

Now
$$\frac{d\psi}{ds} = \frac{d\psi}{d\theta}\frac{d\theta}{ds} = \frac{d\psi}{d\theta}\left(\frac{1}{r}\sin\phi\right) = \frac{d\psi}{d\theta}\frac{1}{r}\frac{1}{\left\{1 + \frac{1}{r^2}\left(\frac{dr}{d\theta}\right)^2\right\}^{\frac{1}{2}}}.$$

We deduce that
$$\frac{d\psi}{ds} = \frac{r^2 - r\frac{d^2r}{d\theta^2} + 2\left(\frac{dr}{d\theta}\right)^2}{\left\{r^2 + \left(\frac{dr}{d\theta}\right)^2\right\}^{\frac{3}{2}}}.$$

The curvature of a curve is defined as $\frac{d\psi}{ds}$, and the radius of curvature as $\left|\frac{ds}{d\psi}\right|$. The result may also be obtained by regarding the equations connecting Cartesian and polar coordinates

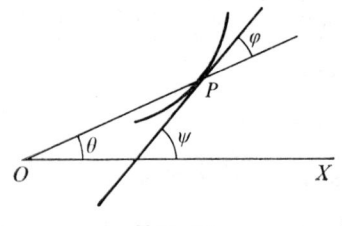

$$x = r\cos\theta, \quad y = r\sin\theta,$$

as a parametric representation of the curve with θ as parameter, and using the result for curvature

$$\frac{\dot{x}\ddot{y} - \ddot{x}\dot{y}}{(\dot{x}^2 + \dot{y}^2)^{\frac{3}{2}}},$$

where dots denote differentiation with respect to θ. From the results

$$\dot{x} = \dot{r}\cos\theta - r\sin\theta, \quad \dot{y} = \dot{r}\sin\theta + r\cos\theta,$$
$$\ddot{x} = \ddot{r}\cos\theta - 2\dot{r}\sin\theta - r\cos\theta,$$
$$\ddot{y} = \ddot{r}\sin\theta + 2\dot{r}\cos\theta - r\sin\theta,$$

we see that
$$\dot{x}\ddot{y}-\ddot{x}\dot{y} = 2\dot{r}^2-r\ddot{r}+r^2,$$
and
$$\dot{x}^2+\dot{y}^2 = r^2+\dot{r}^2,$$

whence the result follows.

The result just proved is rarely used; for many curves in polar form it is simpler to transform the equation into the *p-r* form and use the formula now to be proved.

Radius of curvature in tangential–polar coordinates

If the (p, r) equation of a curve (see Fig. 4.7) is given, then on

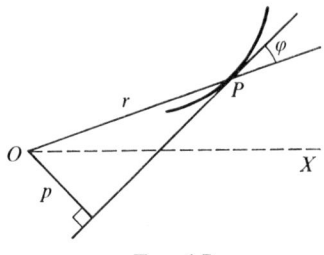

FIG. 4.7

differentiating the equation $p = r \sin \phi$, we have

$$\frac{dp}{dr} = \sin \phi + r \cos \phi \frac{d\phi}{dr}$$

$$= r\frac{d\theta}{ds} + r\left(\frac{dr}{ds}\right)\frac{d\phi}{dr}$$

$$= r\frac{d\theta}{ds} + r\frac{d\phi}{ds}$$

$$= r\frac{d}{ds}(\theta+\phi)$$

$$= r\frac{d}{ds}(\psi)$$

$$= \frac{r}{\rho},$$

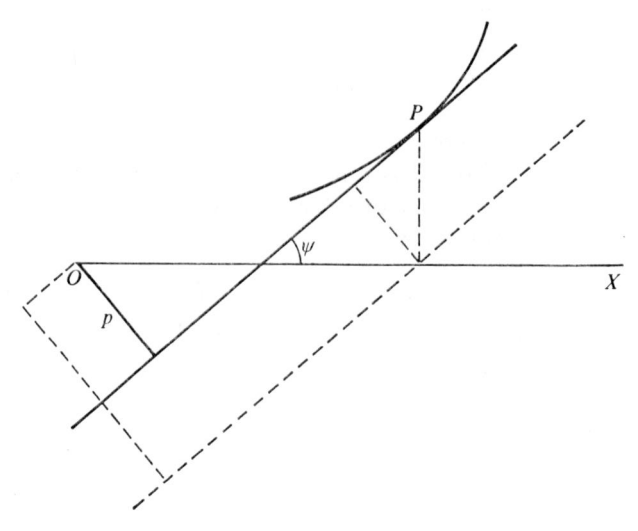

FIG. 4.8

where ρ is the radius of curvature.

$$\therefore \quad \rho = r\frac{dr}{dp}.$$

Example. The *p-r* equation of the cardioid $r = a(1+\cos\theta)$ is $r^3 = 2ap^2$. Thus $3r^2\dfrac{dr}{dp} = 4ap$, and the radius of curvature is

$$\frac{4ap}{3r} = \tfrac{2}{3}\sqrt{(2ar)}.$$

Alternative proof of the formula $\rho = r\dfrac{dr}{dp}$

In Fig. 4.8, if the tangent at $P(x, y)$ has inclination ψ to OX, and p is the length of the perpendicular from O on this tangent,

$$p = x\sin\psi - y\cos\psi.$$

It follows that

$$\frac{dp}{ds} = \left(x\cos\psi\frac{d\psi}{ds} + \sin\psi\frac{dx}{ds}\right) - \left(-y\sin\psi\frac{d\psi}{ds} + \cos\psi\frac{dy}{ds}\right),$$

which, on recalling that $\dfrac{dx}{ds} = \cos \psi$ and $\dfrac{dy}{ds} = \sin \psi$, gives

$$\frac{dp}{ds} = \frac{d\psi}{ds}\left(x\frac{dx}{ds} + y\frac{dy}{ds}\right) = \frac{d\psi}{ds}\frac{1}{2}\frac{d}{ds}(x^2 + y^2)$$

$$= \frac{d\psi}{ds}\frac{1}{2}\frac{d}{ds}(r^2) = \frac{d\psi}{ds} r \frac{dr}{ds}.$$

Hence,
$$\frac{ds}{d\psi} = r\frac{dr}{dp}.$$

Exercises 4(g)

1. Show that the *p-r* equation of the parabola $r(1+\cos \theta) = c$ is $p^2 = cr$ and find the radius of curvature at any point of the parabola in terms of *r*.

2. Find in terms of θ the radius of curvature at any point of the reciprocal spiral $r = a/\theta$.

3. Find the *p-r* equation of the lemniscate $r^2 = a^2 \cos 2\theta$ and hence find the radius of curvature at any point in terms of *r*.

4. Find the radius of curvature of the curve $r^n = a^n \sin n\theta$ at the point $(a, \pi/2n)$.

5. Find the *p-r* equation of the limaçon $r = a(2+\cos \theta)$. Find formulae for the radius of curvature (i) using the polar form, (ii) using the tangential-polar form. Show that the two answers obtained are the same.

6. Prove that the radius of curvature at any point of the rectangular hyperbola $r^2 \cos 2\theta = a^2$ is proportional to the cube of the radius vector.

Some interesting curves

Intrinsic equation of a curve

If *s* denotes the length of the arc *AP* of a curve measured from a fixed point *A* on it, and ψ is the angle made by the tangent at *P* with *OX*, the relation connecting *s* and ψ is called the intrinsic equation of the curve. For some curves the intrinsic equation is the simplest and most natural equation to use. For example, the form of a chain of uniform density, *w* per unit length, suspended between two fixed points is easily found by considering the equilibrium of the portion *AP* of the chain (see Fig. 4.9), where *A* is the lowest point of the chain (at which the tangent to the chain is horizontal).

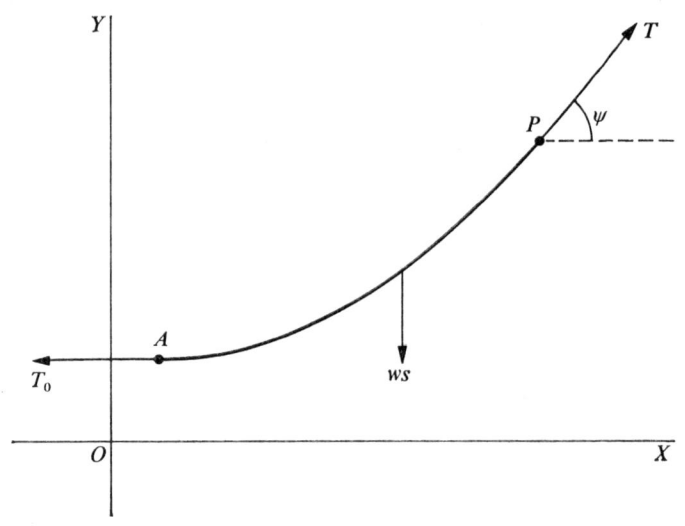

FIG. 4.9

If s denotes the length of the arc AP and ψ the angle made by the tangent at P with the horizontal, then from the equations of equilibrium

$$T \cos \psi = T_0,$$
$$T \sin \psi = ws,$$

we deduce that

$$s = \frac{T_0}{w} \tan \psi,$$

i.e.
$$s = c \tan \psi,$$

where c is a constant. This curve $s = c \tan \psi$ is called a catenary.

The intrinsic equation of the cycloid

$$x = a(\theta + \sin \theta),$$
$$y = a(1 - \cos \theta),$$

is
$$s = 4a \sin \psi.$$

It is not in general easy to sketch a curve from the intrinsic equation; it is however, a simple matter to obtain the Cartesian equation of the curve from the intrinsic equation as follows.

$$\frac{dx}{d\psi} = \frac{dx}{ds} \cdot \frac{ds}{d\psi} = (\cos \psi) \frac{ds}{d\psi},$$
$$\frac{dy}{d\psi} = \frac{dy}{ds} \cdot \frac{ds}{d\psi} = (\sin \psi) \frac{ds}{d\psi}.$$

TECHNIQUES OF INTEGRATION

Now we can find $\dfrac{ds}{d\psi}$ in terms of ψ from the intrinsic equation of the curve. Thus we can find $dx/d\psi$ and $dy/d\psi$ in terms of ψ. On integrating we find x and y each as a function of ψ.

Example. Find the Cartesian equation of the catenary $s = c \tan \psi$.

$$\frac{dx}{d\psi} = \frac{dx}{ds}\frac{ds}{d\psi} = (\cos \psi)\frac{ds}{d\psi} = (\cos \psi)(c \sec^2 \psi)$$
$$= c \sec \psi.$$

$$\frac{dy}{d\psi} = \frac{dy}{ds}\frac{ds}{d\psi} = (\sin \psi)\frac{ds}{d\psi} = (\sin \psi)(c \sec^2 \psi)$$
$$= c \sec \psi \tan \psi.$$

Hence
$$x = \int c \sec \psi \, d\psi = c \log(\sec \psi + \tan \psi) + A,$$

$$y = \int c \sec \psi \tan \psi \, d\psi = c \sec \psi + B.$$

If we choose the origin of the coordinates so that when $\psi = 0$, $x = 0$, $y = c$, then $A = 0$, $B = 0$ and

$$x = c \log(\sec \psi + \tan \psi),$$

$$y = c \sec \psi.$$

These are the parametric equations of the catenary. To obtain the Cartesian equation we eliminate ψ.

$$\sec \psi + \tan \psi = e^{x/c},$$

hence, since $(\sec \psi + \tan \psi)(\sec \psi - \tan \psi) = 1$,

$$\sec \psi - \tan \psi = e^{-x/c}.$$

It follows that
$$2 \sec \psi = e^{x/c} + e^{-x/c} = 2 \cosh \frac{x}{c}.$$

Hence $\sec \psi = \cosh \dfrac{x}{c}$, and it follows that the Cartesian equation of the curve is

$$y = c \cosh \frac{x}{c}.$$

Exercises 4(h)

1. Find the Cartesian equation of the curve whose intrinsic equation is
$$s = a \sec^3 \psi,$$
taking the origin at the point for which $\psi = 0$.

2. Find the Cartesian equation of the curve whose intrinsic equation is
$$s = a \sin^2 \psi,$$
selecting the origin so that when $\psi = 0$, $x = -\frac{2a}{3}$, $y = 0$.

3. The intrinsic equation of a curve is
$$5s = 4a(5 + \tan^2 \tfrac{1}{2}\psi)\sqrt{\tan \tfrac{1}{2}\psi}.$$
If the axes are chosen so that x, y, s, and ψ vanish simultaneously, find the Cartesian coordinates of any point on the curve in terms of ψ and verify that
$$5x^2 + 9y^2 = 5s^2.$$

Curve of pursuit

A dog, speed v_0, pursues a mechanical hare, speed v_1, which moves on a straight line perpendicular to the line joining the original positions of dog and hare. It is required to find the time for the dog to catch the hare.

Let O, N be the original positions of the dog and hare respectively, $ON = a$ (see Fig. 4.10). Then if $P(x, y)$ is the position of the dog

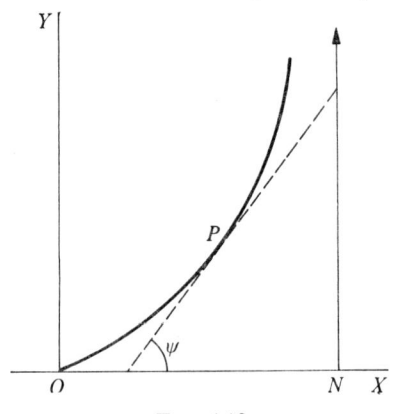

FIG. 4.10

at time t, s the length of the arc OP of its path, and ψ the angle made by the tangent at P with OX,
$$\tan \psi = \frac{v_1 t - y}{a - x} = \frac{\frac{v_1}{v_0} \cdot s - y}{a - x}.$$

TECHNIQUES OF INTEGRATION 109

Write $\dfrac{v_1}{v_0} = k$, then
$$(a-x)\tan \psi = ks - y.$$

On differentiating this equation with respect to x, we have
$$(a-x)\sec^2 \psi \frac{d\psi}{dx} - \tan \psi = k \sec \psi - \tan \psi,$$
$$\therefore (a-x)\sec \psi \frac{d\psi}{dx} = k.$$

Hence
$$\int \sec \psi \, d\psi = k \int \frac{1}{a-x} dx,$$
$$\therefore \log(\sec \psi + \tan \psi) = -k \log(a-x) + A.$$

To find the value of the constant A, we note that when $x = 0$, $\psi = 0$ whence $A = k \log a$. It follows that
$$\log(\sec \psi + \tan \psi) = -k \log\left(\frac{a-x}{a}\right),$$

and hence
$$\sec \psi + \tan \psi = \left(\frac{a-x}{a}\right)^{-k}.$$

Since
$$\sec^2 \psi - \tan^2 \psi = 1,$$

it follows that
$$\sec \psi - \tan \psi = \left(\frac{a-x}{a}\right)^{k},$$
$$\therefore 2 \tan \psi = \left(\frac{a-x}{a}\right)^{-k} - \left(\frac{a-x}{a}\right)^{k}.$$

Now $\tan \psi = \dfrac{dy}{dx}$, hence on integrating the last equation,

$$2y = a^k \frac{(-)(a-x)^{-k+1}}{(-k+1)} + a^{-k} \frac{(a-x)^{k+1}}{(k+1)} + B.$$

Now, when $x = 0$, $y = 0$,
$$\therefore B = \frac{a}{-k+1} - \frac{a}{k+1} = \frac{2ak}{1-k^2}.$$

Hence the Cartesian equation of the curve of pursuit is

$$2y = -a^k\frac{(a-x)^{-k+1}}{-k+1} + a^{-k}\frac{(a-x)^{k+1}}{k+1} + \frac{2ak}{1-k^2}.$$

Now, if $k < 1$, i.e. $v_0 > v_1$, the terms $(a-x)^{-k+1}$ and $(a-x)^{k+1}$ each tends to 0 as x tends to a. Hence the dog will catch the hare when

$$y = \frac{ak}{1-k^2} = \frac{av_0v_1}{v_0^2-v_1^2},$$

provided $v_0 > v_1$; the time taken is $\frac{av_0}{v_0^2-v_1^2}$.

Rolling curves

Suppose two curves are in contact at I, the common tangent at I making an angle ψ with OX (see Fig. 4.11). Let ρ_1 be the radius of

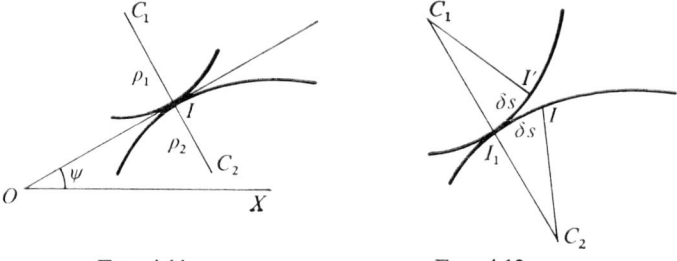

FIG. 4.11   FIG. 4.12

curvature and C_1 the centre of curvature of the upper curve at I, and ρ_2 the radius of curvature and C_2 the centre of curvature of the lower curve at I.

Now suppose the upper curve rolls through a small angle in an anticlockwise direction on the lower curve (see Fig. 4.12), regarded as fixed, so that the new point of contact is I_1, and let I' be the point of the upper curve originally in contact with I. Then, since the centre of curvature is the limiting position of the point of intersection of neighbouring normals, it is clear that

$$I_1 \widehat{C_1} I' = \frac{\delta s}{\rho_1}, \quad I_1 \widehat{C_2} I = \frac{\delta s}{\rho_2}.$$

Thus the angle $\delta\theta$ turned through by the upper curve is given by

$$\delta\theta = \frac{\delta s}{\rho_1} + \frac{\delta s}{\rho_2}.$$

TECHNIQUES OF INTEGRATION   111

This result is important in a problem of stability which we now consider.

Stability

A lamina mass M, centre of mass G, rolls in a vertical plane on a fixed curve and is in equilibrium under the action of M and R, where R is the reaction from the fixed curve (see Fig. 4.13). In this equilibrium position the common tangent to the curves makes an angle ψ with the horizontal, the point of contact of the curves is I, and $IG = h$.

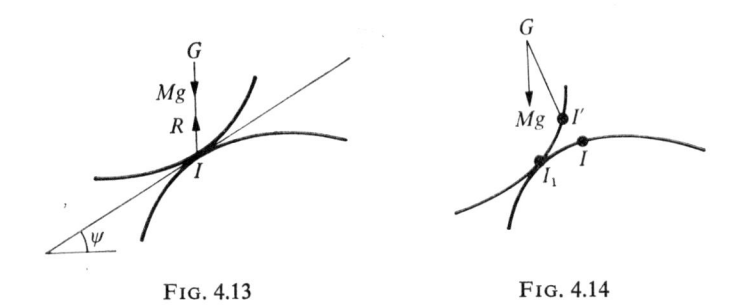

FIG. 4.13   FIG. 4.14

If now, the moving lamina rotates through a small angle $\delta\theta$ in a counter-clockwise direction, so that the new point of contact is I_1, where $II_1 = I'I_1 = \delta s$, then G is displaced $h\,\delta\theta$ horizontally from I' (see Fig. 4.14). But the horizontal component of the displacement $I'I_1$ is $\delta s \cos \psi$. Clearly, if $\delta s \cos \psi > h\,\delta\theta$, the forces acting on the lamina in its new position will have a clockwise moment about I_1 and will tend to rotate the lamina in a clockwise direction and thus restore equilibrium. Thus the position of equilibrium is stable if $\delta s \cos \psi > h\,\delta\theta$, i.e.

$$\delta s \cos \psi > h\left(\frac{\delta s}{\rho_1} + \frac{\delta s}{\rho_2}\right),$$

i.e.

$$\frac{\cos \psi}{h} > \frac{1}{\rho_1} + \frac{1}{\rho_2}.$$

The cycloidal pendulum

A particle is suspended by a string of length l from B, the cusp of a cycloid, so that as the particle oscillates in a vertical plane the string wraps itself alternately round the arcs AB, BC of the cycloid (see Fig. 4.15). We shall find the equation of the locus of the particle Q.

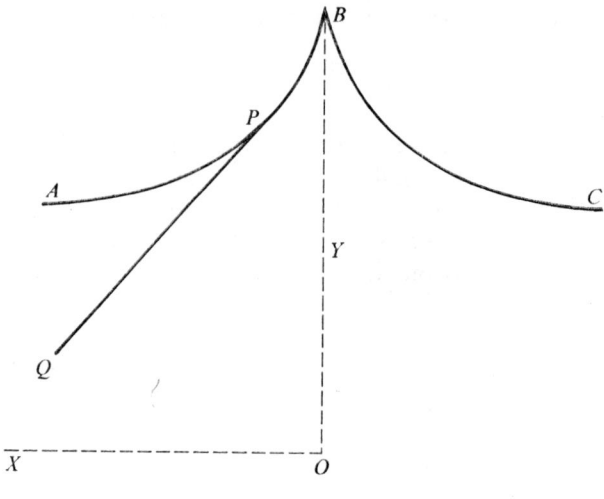

FIG. 4.15

Take as origin the position of the particle when the string is vertical and select horizontal and vertical axes OX, OY as shown in the diagram. Consider the position of Q when the arc of contact of the string with the cycloid is BP, where P is the point

$$\{a(\theta+\sin\theta), a(1-\cos\theta)\},$$

referred to horizontal and vertical axes through A. The arc BP is

$$(4a-4a\sin\tfrac{1}{2}\theta),$$

and hence the length of the straight portion of the string is

$$l-4a+4a\sin\tfrac{1}{2}\theta,$$

and it is inclined to the horizontal at an angle $\tfrac{1}{2}\theta$. Clearly, the coordinates of Q are

$$x = \{a\pi - a(\theta+\sin\theta)\} + (l-4a+4a\sin\tfrac{1}{2}\theta)\cos\tfrac{1}{2}\theta,$$
$$y = l - \{2a - a(1-\cos\theta)\} - (l-4a+4a\sin\tfrac{1}{2}\theta)\sin\tfrac{1}{2}\theta.$$

If we take the length of the string as $4a$, these equations become

$$x = a\pi - a(\theta - \sin\theta),$$
$$y = a(1+\cos\theta);$$

or, on writing $\pi - \theta = \phi$,

$$x = a(\phi + \sin \phi),$$
$$y = a(1 - \cos \phi).$$

Thus the particle describes a cycloidal path. It is easy to show that the particle will perform a simple harmonic oscillation with period $2\pi \sqrt{\dfrac{4a}{g}}$, independent of the amplitude.

A clock based on this principle was constructed by Hughens.

Exercises 4(i)—Parametric equations

1. Sketch the curve whose parametric equations are

$$x = t^4 - 5t^2 + 4, \qquad y = t(1 + t^2).$$

Show how a point on the curve moves as t goes from $-\infty$ to ∞.

2. Sketch the curve

$$x = \frac{2at^2}{1+t^3}, \qquad y = \frac{2at}{1+t^3}.$$

Find the equation of the tangent to the curve at the point with parameter t_1 and prove that this tangent meets the tangent at the point with parameter $-t_1$ in a point on the curve.

3. Sketch the curve whose parametric equations are

$$x = 4t^4 - 37t^2 + 9, \qquad y = t(1 - t^2).$$

Show how a point on the curve moves as t goes from $-\infty$ to ∞.

Miscellaneous problems on Chapter 4

1. Sketch the curve $r = 1 + \cos 2\theta$. Prove that the length of the curve corresponding to $0 \le \theta \le 2\pi$ is $8 + \dfrac{4}{\sqrt{3}} \log(2 + \sqrt{3})$.

2. Find the p-r equation of the hyperbola whose polar equation is $r^2 \cos 2\theta = a^2$. Prove that the radius of curvature at any point of the curve is r^2/p.

3. Find the asymptotes of the curve

$$y^3 - xy^2 - 10x^2y - 8x^3 + 2y^2 + 10xy + 18x^2 - 7y - 2 = 0$$

and sketch the curve.

4. Find the radius of curvature of the curve $s = 4 \sin \psi$ at the point where $\psi = \pi/3$.

5. The ellipse $x^2+4y^2 = 2$ is inverted with respect to a circle centre O, radius 1. Find the r, θ equation and the p-r equation of the inverse curve. Prove that the radius of curvature of the inverse curve where it crosses the x-axis is $1/(2\sqrt{2})$.

6. Prove that ax^2+b can be expressed in the form
$$ax^2+b \equiv p(cx^2+d)+(qx+r)\frac{d}{dx}(cx^2+d),$$
where p, q, r are constants, and find the values of p, q, r in terms of a, b, c, d.

Hence, show that the integral $\int \frac{ax^2+b}{(cx^2+d)^2} dx$ is a rational function of x if and only if $\frac{a}{c}+\frac{b}{d} = 0$.

7. Find the curvature of the curve $s^2 = a^2 \tan^3 \psi$ at the point for which $\psi = \frac{1}{4}\pi$, $s = a$.

8. Sketch the curve $r^2-1 = r \sin \theta$ and find its p-r equation.

9. If $I_n = \int_0^{2a} x^n \sqrt{(2ax-x^2)} \, dx$, show that $I_n = \frac{2n+1}{n+2} aI_{n-1}$ and evaluate I_3.

10. Find the length of the arc of the reciprocal spiral $r = a/\theta$ which lies between $\theta = 1$ and $\theta = 2$.

11. Find a reduction formula connecting $\int \cos m\theta \cos^n \theta \, d\theta$ with
$$\int \cos m\theta \cos^{n-2} \theta \, d\theta.$$

12. Find the asymptotes of the curve
$$2y^3-3x^3+7xy^2+2x^2y-6y^2+11x^2-19xy-4y+10x+4 = 0.$$
Sketch the curve. Find the equation of the tangent at the point $(1, 2)$ on the curve and show that this tangent does not intersect the curve at any other point.

13. Find the Cartesian equation of the curve whose intrinsic equation is
$$s = c(1+\cos^2 \psi),$$
the axes being chosen so that when $\psi = 0$, $x = 2c/3$ and $y = 0$.

14. Obtain a reduction formula for $I_n = \int (x^2+x+1)^{-n} dx$, where n is a positive integer. Hence prove that
$$\int_0^\infty (x^2+x+1)^{-3} dx = \frac{4\pi}{9\sqrt{3}} - \frac{1}{2}.$$

TECHNIQUES OF INTEGRATION

15. Find an expression for the radius of curvature at any point of the curve $r^n = a^n \cos n\theta$.

16. Sketch the region in the third quadrant of the xy-plane defined by the inequalities

$$xy \leq -6, \quad y+x+5 \geq 0, \quad y+2x-1 \leq 0,$$
and
$$x^2+y^2+9x+5y-36 \leq 0,$$

when satisfied simultaneously. Prove that the area of the region is 21·9.

17. A line is drawn through the origin meeting the cardioid $r = a(1+\cos\theta)$ in the points P and Q, and the normals at P and Q intersect in L. Show that the radii of curvature at P and Q are proportional to PL and QL respectively.

18. By integration by parts show that, if $0 < m < n$, and

$$I = \int_0^1 x^m \frac{d^n}{dx^n}\{x^n(1-x)^n\}\, dx,$$

then $I = -m \int_0^1 x^{m-1} \frac{d^{n-1}}{dx^{n-1}}\{x^n(1-x)^n\}\, dx$. Deduce that $I = 0$.

19. Find the asymptote of the curve $x(x^2+3y^2) = 3(x^2-y^2)$. Sketch the curve and find the area of the loop.

20. If $I_{m,n} = \int_0^{\pi/2} \sin^m x \cos 2nx\, dx$, where n is an integer and $m > 1$, prove that

$$(m^2-4n^2)I_{m,n} = m(m-1)I_{m-2,n}.$$

Prove that $I_{1,2} = -\frac{1}{15}$ and use the reduction formula to prove that $I_{5,2} = \frac{8}{63}$.

21. Sketch the curve whose polar equation is $r = \frac{a}{\theta^2-1}$ for values of θ between 1 and 4π. Find the radius of curvature at the point $\theta = 2$. Find also the length of the arc of the spiral from $\theta = 2$ to $\theta = \infty$.

22. Show that, if n is a positive integer, then

$$\int_0^{2\pi} \frac{\cos(n-1)x - \cos nx}{1-\cos x}\, dx = 2\pi,$$

and deduce that
$$\int_0^{2\pi} \left(\frac{\sin \tfrac{1}{2}nx}{\sin \tfrac{1}{2}x}\right)^2 dx = 2n\pi.$$

23. A loop of the curve $r^2 = a^2 \cos 2\theta$ rotates about the line $\theta = \tfrac{1}{2}\pi$. Prove that the area of the surface generated is $2\sqrt{2\pi a^2}$, and find the volume contained by it.

24. If M and N are the feet of the perpendiculars from the pole O on to the tangents at the points P and Q respectively of a given curve, prove that, in the limit, as Q tends to P, the line MN becomes a common tangent to the locus of M and the circle on OP as diameter. If, with the usual notation, the coordinates of P are (p, r), prove that $p^2 = p_1 r$, where p_1 is the length of the perpendicular from O on to this common tangent.

Show that the p-r equation of the curve $r = a \sin^3 \dfrac{\theta}{3}$ is $r^4 = ap^3$, and deduce the p-r equation of the locus of the foot of the perpendicular from the pole on to any tangent to the curve.

25. Show that the asymptotes of the curve
$$x^2 y^2 + xy + x + 2y + 2 = 0$$
are $x = 0$ twice, $y = 0$ twice. Sketch the curve.

26. Sketch the regions of the plane XOY in which the inequality
$$(4x^2 - 3y^2 - 4)(xy - 4) \geq 0 \qquad (1)$$
holds and indicate the points of these areas for which
$$y^2 + 2x = 8. \qquad (2)$$
Show that the set of values of x for points satisfying conditions (1) and (2) simultaneously have no least value and find the greatest value.

27. Prove that all lines of gradients ± 1 meet the curve
$$x^4 - 2x^2 y^2 + y^4 - x^3 - 3x^2 y + y^2 x + 3y^3 - 3x^2 + 6xy + y^2 = 0$$
in two points at infinity, and prove that the equations of the asymptotes are
$$y - x + 1 = 0 \text{ (twice)}; \quad y + x - 1 = 0; \quad y + x + 2 = 0.$$
Prove that the equation of the curve can be written in the form
$$(y - x + 1)^2 (x + y - 1)(x + y + 2) = 5x - 3y - 2.$$
Hence sketch the curve.

28. If $I_n = \int_0^\infty x^{2n} e^{-x^2} dx$, find a relation between I_n and I_{n-1}. Hence express I_n, where n is a positive integer, in terms of I_0.

Hence show that

$$\int_0^\infty \left(\frac{e^{x^2}-1}{x^2}\right) e^{-x^2}\, dx = \left\{1 + \sum_{p=2}^\infty \frac{1.3.5 \cdots (2p-3)}{2^{p-1}(p!)}\right\} I_0.$$

By evaluating the integral on the left-hand side directly in terms of I_0, show that

$$\sum_{p=2}^\infty \frac{1.3.5 \cdots (2p-3)}{2^{p-1}(p!)} = 1.$$

5
Functions of More Than One Variable

So far, we have been concerned with functions of a single variable, but in many problems in science and mathematics we have to deal with functions of two or more independent variables as the following examples illustrate.

(i) The volume v of a given mass of gas is a function of two independent variables: p, the pressure to which the gas is subjected; and T, the temperature of the gas. The law giving v in terms of p and T is $v = k\dfrac{T}{p}$.

(ii) The equation of a certain surface in Cartesian coordinates is $z = x^3 + 2xy$. In this equation we can give x and y any pair of values and calculate the corresponding value of z. z is a function of the two independent variables x and y.

(iii) The lift L of an aeroplane wing is a function of three independent variables: A, the area of the wing; v, the speed at which the wing is moving; and ρ, the density of the air. The law giving L in terms of A, v, and ρ is $L = kAv^2\rho$.

In the language of modern mathematics, the variable z is a function of the two independent variables x and y if there is a correspondence between the set of ordered pairs† (x, y) and a set of real numbers represented by z. The function may be defined for all sets (x, y); e.g. $z = xy$ is such a function. Or the set (x, y) may be restricted to a domain; e.g. if $z = \sqrt{(xy)}$, the domain for which z is defined consists of the first and third quadrants of the xy-plane.

We shall now consider the extension of calculus operations to functions of two or more independent variables. Let $z = f(x, y)$

† The term 'ordered pair' indicates that the order of the two terms x, y is relevant; the fundamental axiom governing ordered pairs is that $(x, y) = (u, v)$ if and only if $x = u$ and $y = v$.

denote a single-valued function of the two independent variables x and y.† The result of differentiating z with respect to x treating y as a constant is called the partial differential coefficient of z with respect to x and is denoted by $\dfrac{\partial z}{\partial x}$; the result of differentiating z with respect to y treating x as a constant is called the partial differential coefficient of z with respect to y and is denoted by $\dfrac{\partial z}{\partial y}$.

Example. If $z = xy+y^2+x^3$, $\dfrac{\partial z}{\partial x} = y+3x^2$ and $\dfrac{\partial z}{\partial y} = x+2y$.

If $z = \log(\sin xy^2)$,

$$\frac{\partial z}{\partial x} = \frac{1}{\sin xy^2} \cos(xy^2) y^2,$$

and

$$\frac{\partial z}{\partial y} = \frac{1}{\sin xy^2} \cos(xy^2) 2xy.$$

A geometrical interpretation of the partial differential coefficients $\dfrac{\partial z}{\partial x}$, $\dfrac{\partial z}{\partial y}$ can be given by considering the surface represented by the equation $z = f(x, y)$ (see Fig. 5.1). Through any point P of the surface we can draw an infinite number of planes perpendicular to the XOY-plane, each cutting the surface in a curve. Two curves of

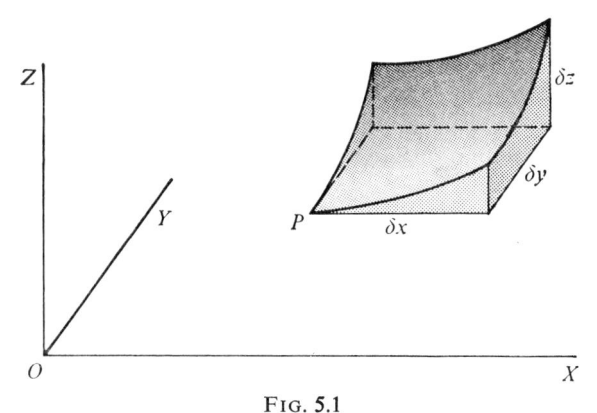

Fig. 5.1

† $z = f(x, y)$ must not be confused with $z = f(xy)$; the latter function is a function of the single variable (xy), e.g. $\log(xy)+(xy)^2$.

section have special importance. On the curve of section by the plane parallel to XOZ, the value of y remains fixed at its value at P while x and z vary. It follows that the gradient of this curve at P is $\dfrac{\partial z}{\partial x}$. Similarly, the gradient of the curve of section by the plane parallel to YOZ is $\dfrac{\partial z}{\partial y}$.

Analytical definition of $\dfrac{\partial z}{\partial x}, \dfrac{\partial z}{\partial y}$

It is clear that, if $z = f(x, y)$,

$$\frac{\partial z}{\partial x} = \lim_{\delta x \to 0} \frac{f(x+\delta x, y) - f(x, y)}{\delta x},$$

$$\frac{\partial z}{\partial y} = \lim_{\delta y \to 0} \frac{f(x, y+\delta y) - f(x, y)}{\delta y}.$$

There is a natural extension of partial differentiation to functions of three or more variables.

Example. If $z = xy^2 + xyt + 2t^4$, where x, y, t are independent variables,

$$\frac{\partial z}{\partial x} = y^2 + yt,$$

$$\frac{\partial z}{\partial y} = 2xy + xt,$$

$$\frac{\partial z}{\partial t} = xy + 8t^3.$$

Exercises 5(a)

1. If $z = xy^3(x^2+y)$, find the values of $\dfrac{\partial z}{\partial x}, \dfrac{\partial z}{\partial y}$. Find the gradient of the section of the surface by the plane through (1, 1, 2) parallel to YOZ. Also find the gradient of the section of the surface by the plane through (2, 1, 10) parallel to XOZ.

2. If $z = e^{x^2+y^2}$, prove that $y\dfrac{\partial z}{\partial x} = x\dfrac{\partial z}{\partial y}$.

3. If $z = \sin(x+2y)$, find the value of $\dfrac{\partial z}{\partial x}$ when $x = \tfrac{1}{4}\pi$, $y = \tfrac{1}{4}\pi$, and the value of $\dfrac{\partial z}{\partial y}$ when $x = 0$, $y = \tfrac{1}{2}\pi$.

FUNCTIONS OF MORE THAN ONE VARIABLE

4. If $z = \log(x+y^2+xyt^2)$, where x, y, t are independent variables, find the values of $\dfrac{\partial z}{\partial x}, \dfrac{\partial z}{\partial y}, \dfrac{\partial z}{\partial t}$ when $x = 2, y = 1, t = 1$.

5. The centre of a conic $f(x, y) \equiv ax^2+2hxy+by^2+2gx+2fy+c = 0$ is found by transferring the origin to (h, k) and selecting h and k so that the coefficients of the first degree terms in x and y vanish in the new equation. Find the two equations giving the centre and show that $x = h$, $y = k$, are solutions of the equations $\dfrac{\partial f}{\partial x} = 0, \dfrac{\partial f}{\partial y} = 0$.

6. Find the values of $\dfrac{\partial z}{\partial x}$ and $\dfrac{\partial z}{\partial y}$ at the point $(2, 1, 3)$ on the sphere $x^2+y^2+z^2 = 14$.

7. In a rectilinear wave motion along a straight line OX, each particle of the medium oscillates with simple harmonic motion in a direction perpendicular to OX, the phase of the S.H.M. being dependent on the x-coordinate of the particle. The displacement y of the particle in a direction perpendicular to OX is given by

$$y = a\cos(bt+cx),$$

where a, b, c are constants.

Thus, y is a function of two independent variables t and x; but for any one particle, the value of x is fixed. State an expression for the velocity of a particle and find the value of the velocity in terms of t and x.

8. If $u = f(x, y, z)$, where x, y, z are independent variables, state analytical definitions of $\dfrac{\partial u}{\partial x}, \dfrac{\partial u}{\partial y}, \dfrac{\partial u}{\partial z}$.

Potential of a field

An important function in mathematical physics is the *potential* of a gravitational, magnetic, or electrostatic field which is defined at any point P of the field as the negative of the work done by the forces of the field as a unit particle on which these forces act moves to P from a standard position.

Consider a particle moving in a plane XOY and suppose that in any position (x, y) the components of the forces of the field parallel to OX, OY are P, Q respectively, where P and Q are functions of x and y (see Fig. 5.2). Then the work done in moving from (x, y) to $(x+\delta x, y+\delta y)$ is, to the first order of small quantities, $P\,\delta x + Q\,\delta y$.

Hence if V, a function of x and y, is the potential at (x, y),

$$\delta V = -(P\,\delta x + Q\,\delta y).$$

It follows that $\dfrac{\partial V}{\partial x} = -P, \dfrac{\partial V}{\partial y} = -Q.$

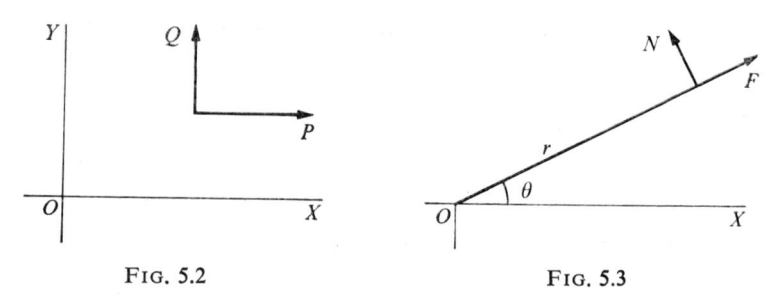

FIG. 5.2

FIG. 5.3

If the position of the particle is given in polar coordinates (r, θ) (see Fig. 5.3) and the component of the forces of the field along and perpendicular to the radius vector are F, N respectively where F and N are functions of r and θ, then

$$\delta V = -(F\,\delta r + Nr\,\delta\theta).$$

It follows that $\dfrac{\partial V}{\partial r} = -F$, $\dfrac{\partial V}{\partial \theta} = -Nr$.

Higher order partial differential coefficients

If z is a function of x and y, then $\dfrac{\partial z}{\partial x}$ is also a function of x and y and may be differentiated partially with respect to x and partially with respect to y. $\dfrac{\partial}{\partial x}\left(\dfrac{\partial z}{\partial x}\right)$ is denoted by $\dfrac{\partial^2 z}{\partial x^2}$, and $\dfrac{\partial}{\partial y}\left(\dfrac{\partial z}{\partial x}\right)$ is denoted by $\dfrac{\partial^2 z}{\partial y\,\partial x}$.

Similarly, we can differentiate $\dfrac{\partial z}{\partial y}$ partially with respect to x and partially with respect to y. $\dfrac{\partial}{\partial x}\left(\dfrac{\partial z}{\partial y}\right)$ is denoted by $\dfrac{\partial^2 z}{\partial x\,\partial y}$, and $\dfrac{\partial}{\partial y}\left(\dfrac{\partial z}{\partial y}\right)$ is denoted by $\dfrac{\partial^2 z}{\partial y^2}$.

Example. If $z = x^2 y^2 + x^3 + 4y$,

$$\frac{\partial z}{\partial x} = 2xy^2 + 3x^2,$$

whence

$$\frac{\partial^2 z}{\partial x^2} = 2y^2 + 6x,$$

and
$$\frac{\partial^2 z}{\partial y \, \partial x} = 4xy.$$

Also
$$\frac{\partial z}{\partial y} = 2x^2 y + 4,$$

whence
$$\frac{\partial^2 z}{\partial x \, \partial y} = 4xy,$$

and
$$\frac{\partial^2 z}{\partial y^2} = 2x^2.$$

It will be noticed in the above results that $\dfrac{\partial^2 z}{\partial y \, \partial x} = \dfrac{\partial^2 z}{\partial x \, \partial y}$. Discussion of the type of function for which this result is valid is beyond the scope of this volume; but we can say that while it is possible for mathematicians to invent recondite functions for which

$$\frac{\partial^2 z}{\partial y \, \partial x} \neq \frac{\partial^2 z}{\partial x \, \partial y},$$

the functions we normally deal with are well behaved in this respect and satisfy
$$\frac{\partial^2 z}{\partial y \, \partial x} = \frac{\partial^2 z}{\partial x \, \partial y}.$$

Throughout the chapter, we shall assume this property for all functions under consideration.

Exercises 5(a) (cont.)

9. If $z = x^3 y + xy^2 + 3y$, find the values of $\dfrac{\partial^2 z}{\partial x^2}$, $\dfrac{\partial^2 z}{\partial x \, \partial y}$, $\dfrac{\partial^2 z}{\partial y^2}$, when $x = 1$, $y = -1$.

10. If $z = \log(x^3 + y^3)$, find the values of $\dfrac{\partial^2 z}{\partial x^2}$, $\dfrac{\partial^2 z}{\partial y^2}$, when $x = 1$, $y = 2$.

11. If $u = 2xy + \left(\dfrac{y}{x}\right)^4$, prove that
$$x^2 \frac{\partial^2 u}{\partial x^2} + 2xy \frac{\partial^2 u}{\partial x \, \partial y} + y^2 \frac{\partial^2 u}{\partial y^2} = 4xy.$$

12. If $z = \log(x^2 + y^2)$, prove that $\dfrac{\partial^2 z}{\partial x^2} + \dfrac{\partial^2 z}{\partial y^2} = 0$.

13. If $z = xe^{2y} + ye^{2x}$, prove that $\dfrac{\partial^2 z}{\partial x^2} + \dfrac{\partial^2 z}{\partial y^2} = 4z$.

14. If $z = x^2 e^{xy^2}$, prove that $\dfrac{\partial}{\partial x}\left(\dfrac{\partial z}{\partial y}\right) = \dfrac{\partial}{\partial y}\left(\dfrac{\partial z}{\partial x}\right)$.

Maximum and minimum values of a function of two variables

By considering the surface represented by the equation $z = f(x, y)$, it is clear that if z has a maximum (or minimum) value at P, all the curves of section of the surface by planes through P parallel to OZ will also have a maximum (or minimum) value at P. In particular, the two principal sections have a maximum (or minimum) value at P. It follows that the conditions

$$\frac{\partial z}{\partial x} = 0, \quad \frac{\partial z}{\partial y} = 0$$

are *necessary* conditions for z to have a maximum or minimum value. It will also be clear that these conditions are not sufficient; for example, with these conditions satisfied, the surface could have a saddle-point at P with one principal section possessing a maximum at P while the other principal section possesses a minimum at P. In practical problems, it is usually easy to decide by examining values of z in the region of the point whether a particular solution of the equations $\partial z/\partial x = 0$, $\partial z/\partial y = 0$ gives a maximum or a minimum or neither. Other methods of distinguishing between the solutions will be given later.

Example. $z = x^3 + y^3 - 3(x+y)^2$.

$$\frac{\partial z}{\partial x} = 3x^2 - 6(x+y), \quad \frac{\partial z}{\partial y} = 3y^2 - 6(x+y).$$

On putting $\frac{\partial z}{\partial x} = 0$, $\frac{\partial z}{\partial y} = 0$, we obtain two solutions: $x = 4$, $y = 4$; $x = 0$, $y = 0$. Thus there are possible maximum or minimum values of z at $P(4, 4, -64)$ and $O(0, 0, 0)$.

Consider the points on the surface in the neighbourhood of P. If $(4+h, 4+k, -64+\delta z)$ is a neighbouring point, where h and k are small, then

$$-64 + \delta z = (4+h)^3 + (4+k)^3 - 3(8+h+k)^2,$$

whence

$$\delta z = 9h^2 + 9k^2 - 6hk + \ldots$$
$$= 3(h-k)^2 + 6h^2 + 6k^2 + \text{terms of third order in } h \text{ and } k.$$

Thus, for all small h and k, the sign of δz is governed by the sign of $3(h-k)^2 + 6h^2 + 6k^2$, which is positive. It follows that P is a minimum point of the surface. Thus when $x = 4$, $y = 4$, z has a minimum value -64.

Applying a similar procedure for the point O, we find that the sign of δz is governed by the sign of $-3(h+k)^2$, which is negative for all small h and k. It follows that O is a maximum point of the surface. Thus when $x = 0$, $y = 0$, z has a maximum value 0.

Line of best fit

Many laws of science connecting a pair of experimental variables x and y fall into one or other of the following classes:

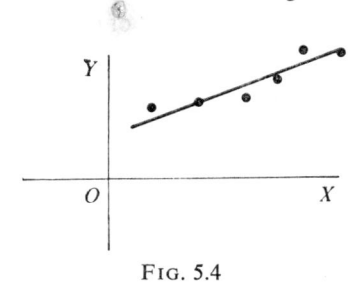

FIG. 5.4

(i) $y = ax+b$. For example, y, the length of a solid bar, is a function of x, the temperature of the bar, the relation between x and y is $y = ax+b$.

(ii) $y = cx^k$. For example, y, the period of a simple pendulum of length x, is given by $y = cx^{\frac{1}{2}}$.

To test these laws and find the values of the constants we usually employ a graphical method (see Fig. 5.4). We find by experiment the values of y corresponding to a set of values of x; in case (i) we graph y against x, in case (ii) we graph $\log y$ against $\log x$. The set of points thus obtained from our experimental results should lie on a straight line. In practice, we are likely, because of experimental error, to obtain a set of points which do not lie perfectly but only approximately on a straight line, as illustrated in Fig. 5.4. The question then arises of finding the straight line which best fits the experimental data. With, say, ten experimental results we can attempt to draw by eye a line which lies evenly with respect to the set of points. But this is an unreliable method, especially if we have a large number of experimental results, and so it is desirable to devise a technique for obtaining the line of best fit. The technique adopted is to calculate S, the sum of the squares of the distances of the points from a line

$y = mx+c$, and then choose the values of m and c so that S is a minimum. This method is called the *principle of least squares* and the line obtained is called the line of best fit.

$$S = \sum \frac{(y-mx-c)^2}{1+m^2}.$$

Let N be the total number of points. We select the origin of coordinates to be at the mean centre of the distribution, so that $\sum x = 0$, $\sum y = 0$. Let σ_1 be the standard deviation of the x's, defined by $N\sigma_1^2 = \sum x^2$. Let σ_2 be the standard deviation of the y's, defined by $N\sigma_2^2 = \sum y^2$. We denote the coefficient of correlation between the x's and the y's by r; this is defined by

$$r = \frac{1}{N\sigma_1\sigma_2} \sum xy.$$

Thus $\sum x = 0$, $\sum y = 0$, $\sum x^2 = N\sigma_1^2$, $\sum y^2 = N\sigma_2^2$, $\sum xy = Nr\sigma_1\sigma_2$. It follows that

$$S = N\left\{\frac{c^2+\sigma_2^2+m^2\sigma_1^2-2mr\sigma_1\sigma_2}{1+m^2}\right\}.$$

The maximum and minimum values of S occur when $\frac{\partial S}{\partial m} = 0$, $\frac{\partial S}{\partial c} = 0$. $\frac{\partial S}{\partial m} = 0$ gives

$$(1+m^2)(2m\sigma_1^2-2r\sigma_1\sigma_2)-(c^2+\sigma_2^2+m^2\sigma_1^2-2mr\sigma_1\sigma_2)(2m) = 0.$$

$\frac{\partial S}{\partial c} = 0$ gives $c = 0$.

It follows that the line of best fit passes through the mean centre of the distribution, its gradient being given by

$$(1+m^2)(2m\sigma_1^2-2r\sigma_1\sigma_2)-(\sigma_2^2+m^2\sigma_1^2-2mr\sigma_1\sigma_2)(2m) = 0,$$

i.e.
$$(1-m^2)r\sigma_1\sigma_2 = m(\sigma_1^2-\sigma_2^2).$$

In practice, there is no difficulty in deciding which root of this quadratic equation in m to take for the line of best fit.

Example. For the set of seven points (5, 2), (10, 5), (12, 6), (18, 10), (27, 13), (31, 16), (37, 18), $\bar{x} = 20$, $\bar{y} = 10$, $7\sigma_1^2 = 852$, $7\sigma_2^2 = 214$, and $r\sigma_1\sigma_2 = 425$. Hence the equation giving the gradient of the line of best fit is

$$638m = 425(1-m^2),$$

whence the required gradient is
$$\frac{-638+\sqrt{((638)^2+4(425)^2)}}{850} = \frac{1}{2}.$$
Hence the equation of the line of best fit is
$$y-10 = \tfrac{1}{2}(x-20).$$

Exercises 5(b)

1. Show that the line of best fit for the following set of points (2, 2), (4, 4), (6, 7), (8, 12), (10, 13), (12, 18), (14, 21) is $y = 1{\cdot}64x - 2{\cdot}12$.

2. Consider the stationary points of the following surfaces:
   (i) $z = x^3 - 3x^2 - y^2 + 6$,
   (ii) $z = y^3 - 6x^2y$.

3. For the surface $z = x^4 + y^4 - 2(x-y)^2$, prove that z has a minimum value when $x = \sqrt{2}$, $y = -\sqrt{2}$, and also when $x = -\sqrt{2}$, $y = \sqrt{2}$. What shape has the surface at the origin?

4. If we have a set of N pairs of values of two statistically connected variables, a typical pair being (x, y) relative to the mean centre of the distribution as origin, we define the *line of regression* of the y's on the x's as the line $y = mx + c$ for which
$$s = \sum (y - mx - c)^2$$
is a minimum. Prove that the equation of this regression line is
$$y = \frac{r\sigma_2 x}{\sigma_1},$$
where r is the coefficient of correlation.

Similarly the line of regression of the x's on the y's is defined as the line $x = ty + d$ for which
$$v = \sum (x - ty - d)^2$$
is a minimum. Find the equation of this regression line, and show that the two regression lines coincide if the coefficient of correlation is unity. Show that in this case the regression lines coincide with the line of best fit of the distribution.

5. A triangle ABC has sides a, b, c and area Δ. P is any point within the triangle and x, y, z are the lengths of the perpendiculars from P onto the sides BC, CA, AB. If u denotes the sum of the squares of the perpendiculars from P to the sides of the triangle, show that u may be written in the form
$$u = x^2 + y^2 + \left\{\frac{2\Delta - ax - by}{c}\right\}^2,$$

where x and y are independent variables. Hence, find the minimum value of u.

6. Determine any maximum or minimum values of the functions

   (i) $(x+y-6)(x^2+y^2)$;   (ii) $(x+y-1)(x^2+y^2)$.

7. Find the values of x and y for which $z = (x^2+2y^2)e^{-x^2-y^2}$ has stationary values. Hence investigate the maximum and minimum values of the function.

Envelope of a family of curves

The equation of the tangent at $(at^2, 2at)$ to the parabola $y^2 = 4ax$ is $yt = x+at^2$; two neighbouring members of the family of tangents, $yt_1 = x+at_1^2$ and $yt_2 = x+at_2^2$ intersect at $(at_1 t_2, a(t_1+t_2))$ and, as t_2 tends to t_1, this point of intersection tends to $(at_1^2, 2at_1)$, a point on the parabola. Thus we may regard the parabola as the locus of the limiting position of the point of intersection of neighbouring members of the family of tangents, and call the parabola the envelope of the family of straight lines.

The envelope of a family of circles of radius a with their centres on the circumference of a circle of radius $2a$ consists of two circular arcs, one of radius a and the other of radius $3a$.

In general, the envelope of a family of curves is defined as the locus of the limiting position(s) of the point(s) of intersection of neighbouring members of the family.

The family of lines considered above has equation $yt = x+at^2$, where t is a parameter fixed on any one line of the system, but varying from line to line; such a parameter is called a variable parameter. The family of circles considered in the second example has equation

$$(x-2a\cos t)^2 + (y-2a\sin t)^2 = a^2,$$

where t is a variable parameter.

We shall consider the envelope of a family of curves $f(x, y, t) = 0$, where t is a variable parameter, i.e. a parameter fixed on any one curve of the system but varying from curve to curve. Consider two neighbouring curves of the system,

$$f(x, y, t) = 0, \tag{1}$$

and

$$f(x, y, t+\delta t) = 0. \tag{2}$$

The point of intersection of these neighbouring members of the family is given by solving eqns (1) and (2) or any pair of independent equations obtained from (1) and (2). Thus, the point of intersection is given by eqn (1) and

$$\frac{f(x, y, t+\delta t) - f(x, y, t)}{\delta t} = 0. \tag{3}$$

It follows that the limiting position as δt tends to 0 of the point of intersection of neighbouring members of the family is given by the equations

$$f(x, y, t) = 0 \quad \text{and} \quad \frac{\partial}{\partial t} f(x, y, t) = 0.$$

The Cartesian equation of the envelope of the family (1) is found by eliminating t from these two equations.

If the equation of the family is given in polar coordinates, $f(r, \theta, t) = 0$, where t is a variable parameter, the polar equation of the envelope is found by eliminating t from the equations

$$f(r, \theta, t) = 0 \quad \text{and} \quad \frac{\partial}{\partial t} f(r, \theta, t) = 0.$$

Example. Consider the family of lines of constant length a, having their extremities one on each of two perpendicular lines OX, OY (see Fig. 5.5). The equation of a member of the family is

$$\frac{x}{a \cos \theta} + \frac{y}{a \sin \theta} = 1$$

and the family is represented by this equation, where θ is a variable parameter. It follows that the envelope of the family is found by

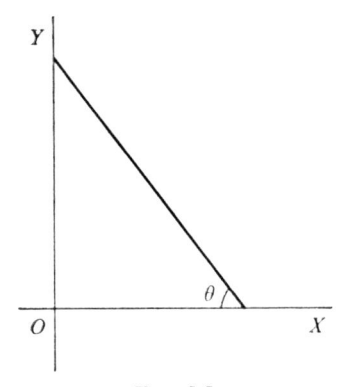

FIG. 5.5

eliminating θ from the equations

$$\frac{x}{a\cos\theta}+\frac{y}{a\sin\theta}=1 \quad \text{and} \quad \frac{x}{a}\left(\frac{\sin\theta}{\cos^2\theta}\right)+\frac{y}{a}\left(\frac{-\cos\theta}{\sin^2\theta}\right)=0,$$

which give $x = a\cos^3\theta$, $y = a\sin^3\theta$, the parametric equations of the astroid

$$x^{\frac{2}{3}}+y^{\frac{2}{3}}=a^{\frac{2}{3}}.$$

Exercises 5(c)

1. Prove that the envelope of a family of lines, each of which forms with OX, OY a triangle of constant area k^2, is the hyperbola $2xy = k^2$.

2. If a particle is projected from O with a velocity v at inclination θ to the horizontal, the path of the particle, referred to horizontal and vertical axes through O, is the parabola

$$y = x\tan\theta - \frac{gx^2}{2v^2}\sec^2\theta.$$

Prove that the envelope of the family of paths obtained by keeping the velocity of projection constant but varying θ is the parabola

$$x^2 = \frac{v^2}{g^2}(v^2 - 2gy).$$

Deduce the maximum possible range of the particle for a given velocity v (i) on a horizontal plane (ii) on a plane inclined at $\frac{1}{4}\pi$ to the horizontal.

3. Find the envelope of the family of lines $y = tx + \sqrt{(a^2t^2+b^2)}$ where t is a variable parameter.

4. From a fixed point O, a line is drawn to intersect a fixed line AB in P. From P a line PQ is drawn perpendicular to OP. Show that as P varies, the family of lines PQ envelope a parabola. Carry out the construction and obtain the envelope practically.

5. P is the point ($a\cos\theta$, $a\sin\theta$) on the fixed circle $x^2+y^2 = a^2$, and C is the fixed point (k, 0). Find the equation of the line through P perpendicular to CP and find the envelope of the family of lines formed as θ varies. Prove that the envelope is an ellipse or hyperbola according as k is less or greater than a. Also construct the envelopes graphically when $k = \frac{3}{4}a$, $k = \frac{3}{2}a$.

6. Find the equation of the focal chord, gradient m, of the parabola $y^2 = 4ax$, and the equation of the circle having this focal chord as a diameter. Show that the envelope of the family of circles obtained by varying m is the curve

$$x^3+xy^2-2ax^2+ay^2-3a^2x = 0.$$

Also construct the envelope graphically.

FUNCTIONS OF MORE THAN ONE VARIABLE

7. P is the point with coordinates $(c\cos\theta, c\sin\theta)$ on the circle $x^2+y^2=c^2$. Q is the point with coordinates $(c\cos(\theta+\alpha), c\sin(\theta+\alpha))$. Find the equation of the chord PQ. Find the equation of the envelope of the family of lines PQ as θ varies, α remaining constant, and show that the envelope is a circle.

8. O is a fixed point on a circle of centre C and radius a. OP is a chord of the circle with angle $\angle COP = \phi$. Taking O as pole and OC as initial line, find the polar equation of the circle on OP as diameter, and find the polar equation of the envelope of the family of circles as ϕ varies. Prove that the envelope is a cardioid.

9. Construct the envelope of a fixed diameter of a circle radius a which rolls on a fixed straight line. Prove theoretically that the envelope is a cycloid.

10. A system of ellipses has the same principal axes and fixed area $k^2\pi$. Find the envelope of the family.

Evolute of a curve

The evolute of a curve is defined as the locus of the centre of curvature and it is easily seen that the centre of curvature is the limiting position of the point of intersection of neighbouring normals. It follows that the evolute of a curve is the envelope of the family of normals to the curve.

Example. For the parabola $y^2 = 4ax$, the equation of the normal, gradient m, is

$$y = mx - 2am - am^3. \qquad (1)$$

The envelope of the normals is given by eliminating m between (1) and

$$0 = x - 2a - 3am^2,$$

and hence the equation of the evolute of the parabola is

$$27ay^2 = 4(x-2a)^3.$$

The gradients m of the normals to the parabola that can be drawn from a given point (x_1, y_1) are given by the cubic equation

$$y_1 = mx_1 - 2am - am^3,$$

i.e.

$$m^3 - \left(\frac{x_1-2a}{a}\right)m + \frac{y_1}{a} = 0.$$

Now it is well known† that the cubic equation $m^3 - pm + q = 0$ has

† Page, A., *Algebra*, p. 65. U.L.P.

three real roots if $4p^3 \geq 27q^2$ and one real root if $4p^3 < 27q^2$. Hence, it is possible to draw three normals to the parabola from (x_1, y_1) if

$$4(x_1-2a)^3 \geq 27ay_1^2,$$

and only one normal if

$$4(x_1-2a)^3 < 27ay_1^2.$$

Thus the evolute of a parabola marks the boundary between those points from which three normals can be drawn to the parabola and those points for which only one normal can be drawn.

Exercises 5(c) (cont.)

11. Find the equation of the normal to the cycloid

$$x = a(t+\sin t), \qquad y = a(1-\cos t)$$

and prove that the equation of the evolute of the cycloid may be written in the parametric form

$$x = a(t-\sin t), \qquad y-2a = a(1+\cos t),$$

another cycloid of the same size as the given cycloid. Sketch one arc of the cycloid and its evolute.

12. Prove that the evolute of the ellipse

$$\frac{x^2}{a^2}+\frac{y^2}{b^2} = 1$$

is the curve

$$(ax)^{\frac{2}{3}}+(by)^{\frac{2}{3}} = (a^2-b^2)^{\frac{2}{3}}.$$

13. Find the equation of the evolute of the hyperbola $\dfrac{x^2}{a^2}-\dfrac{y^2}{b^2} = 1$.

The epicycloid as an envelope

The equation of the line joining the points $(c \cos \theta, c \sin \theta)$ and $(c \cos k\theta, c \sin k\theta)$ on the circle $x^2+y^2 = c^2$ is

$$x \cos\{\tfrac{1}{2}(1+k)\theta\}+y \sin\{\tfrac{1}{2}(1+k)\theta\} = c \cos\{\tfrac{1}{2}(1-k)\theta\}. \qquad (1)$$

The envelope of this family of lines as θ varies is given by (1) and

$$-x\tfrac{1}{2}(1+k)\sin\{\tfrac{1}{2}(1+k)\theta\}+y\tfrac{1}{2}(1+k)\cos\{\tfrac{1}{2}(1+k)\theta\}$$
$$= -c\tfrac{1}{2}(1-k)\sin\{\tfrac{1}{2}(1-k)\theta\}. \qquad (2)$$

FUNCTIONS OF MORE THAN ONE VARIABLE 133

Solving eqns (1) and (2) for x and y, we have

$$x = \frac{c}{1+k}(k \cos \theta + \cos k\theta),$$

$$y = \frac{c}{1+k}(k \sin \theta + \sin k\theta).$$

These are the parametric equations of the envelope of the family of lines (1).

Now the locus of a point on the circumference of a circle of radius b which rolls on the outside of a fixed circle of radius a is a curve with parametric equations

$$x = (a+b)\cos t + b \cos\left(\frac{a+b}{b}\right)t,$$

$$y = (a+b)\sin t + b \sin\left(\frac{a+b}{b}\right)t.$$

The curve is called an epicycloid. On comparing the parametric equations of the envelope of the family of lines (1) with the standard parametric equations of the epicycloid, it is clear that the two pairs of equations are the same if we take $k = (a+b)/b$ and $c = a+2b$. Thus we can construct an epicycloid as the envelope of a family of chords of a circle. In order to construct an epicycloid with n cusps, we must take $k = 1+n$.

Exercises 5(d)

1. A circle of radius 2 cm rolls on the outside of a circle of radius 6 cm. Sketch the epicycloid traced by a point on the rolling circle. Now construct a circle of radius 10 cm and draw a set of chords joining points (10 cos θ, 10 sin θ) and (10 cos 4θ, 10 sin 4θ). Show that this set of chords envelopes the epicycloid.

2. Draw a set of chords joining the points (4 cos θ, 4 sin θ) and (4 cos 3θ, 4 sin 3θ) on the circle $x^2+y^2 = 16$. Sketch the curve enveloped by the chords.

3. Find the equation of the line joining the points ($c \cos \theta$, $c \sin \theta$) and ($c \cos(-k\theta)$, $c \sin(-k\theta)$) on the circle $x^2+y^2 = c^2$. Prove that, as θ varies, the family of lines envelopes an hypocycloid. What is the appropriate value of k to construct a hypocycloid with n cusps?

4. Construct a three-cusped hypocycloid as an envelope of chords of a circle.

5. Construct a four-cusped hypocycloid as an envelope of chords of a circle.

Notation for partial differential coefficients

Let $z = f(x, y)$, where x and y are independent variables. We have so far denoted the partial differential coefficient of z with respect to x by $\dfrac{\partial z}{\partial x}$; it may also be denoted by $\dfrac{\partial f}{\partial x}$.

If $z = f(x, y, t)$, where t is a function of x and y, where x and y are independent variables, then z may be regarded either as a function of x, y, t or as a function of x and y only. In cases like this we distinguish between $\partial z/\partial x$ and $\partial f/\partial x$. By $\dfrac{\partial z}{\partial x}$ we shall understand the partial differential coefficient of z with respect to x when z is expressed as a function of x and y only; by $\dfrac{\partial f}{\partial x}$ we shall understand the partial differential coefficient of $f(x, y, t)$ with respect to x, treating both y and t as constants. Thus $\partial z/\partial x$ and $\partial f/\partial x$ have different meanings.

Example. $z = f(x, y, t) \equiv x^2 + y^3 + txy$, where $t = xy^2$. Then

$$\frac{\partial f}{\partial x} = 2x + ty = 2x + xy^3.$$

On expressing z as a function of x and y only,

$$z = x^2 + y^3 + x^2y^3 \quad \text{and} \quad \frac{\partial z}{\partial x} = 2x + 2xy^3,$$

which differs from $\dfrac{\partial f}{\partial x}$.

If $w = f(x, y, z, u, v)$, where u and v are functions of the three independent variables x, y, z, then $\partial f/\partial x$ means the partial differential coefficient of $f(x, y, z, u, v)$ with respect to x, treating y, z, u, v as constants: but $\partial w/\partial x$ means the partial differential coefficient of z with respect to x when w is expressed as a function of x, y, z only.

If $V = xz + yu$, where $x + y + z = 1$, $x + 2y + u = 1$, then V may variously be regarded as a function of two independent variables which may be x and y, or x and z, or x and u, or y and u, etc., and we shall obtain different answers for the partial differential coefficients according to the selection of the independent variables. Thus, it is clearly necessary, in evaluating partial differential

FUNCTIONS OF MORE THAN ONE VARIABLE 135

coefficients to state, if there is any ambiguity, which variables are regarded as the fundamental independent variables.

In the case considered, we find, if we express V as a function of x and y only, that

$$\frac{\partial V}{\partial x} = 1-2x-2y,$$

and to make it clear what we are regarding as the independent variables, we can write

$$\left(\frac{\partial V}{\partial x}\right)_y = 1-2x-2y.$$

If, on the other hand, we express V as a function of x and u only, we find

$$\left(\frac{\partial V}{\partial x}\right)_u = \tfrac{1}{2}-x.$$

Exercises 5(e)

1. If $z = f(x, y, t) \equiv xyt^2$, where $t = x^2+y^2$, find $\dfrac{\partial z}{\partial x}$ and $\dfrac{\partial f}{\partial x}$.

2. If $z = f(x, y, t) \equiv x^2y^3t+3t^2$, where $t = 2x+y^4$, find the values when $x = 1$, $y = 1$ of $\dfrac{\partial z}{\partial x}, \dfrac{\partial z}{\partial y}, \dfrac{\partial f}{\partial x}, \dfrac{\partial f}{\partial y}$.

3. If $w = \phi(x, y, u, v) \equiv x^2u+y^2v+uv$, where $u = xy$, $v = x+y$, find $\dfrac{\partial w}{\partial x}, \dfrac{\partial w}{\partial y}, \dfrac{\partial \phi}{\partial x}, \dfrac{\partial \phi}{\partial y}$.

4. If $t = x^2u+y^2v$, where $x+y+u = 0$, $2x+u+v = 0$, find the values when $x = 1$, $y = 2$ of $\left(\dfrac{\partial t}{\partial y}\right)_x, \left(\dfrac{\partial t}{\partial y}\right)_u, \left(\dfrac{\partial t}{\partial y}\right)_v, \left(\dfrac{\partial t}{\partial x}\right)_u$.

Small changes in variables

If $z = f(x, y)$, where x and y are independent variables, we shall investigate a formula connecting corresponding small increments $\delta x, \delta y, \delta z$ in the values of x, y, z.

Since $z+\delta z = f(x+\delta x, y+\delta y)$, it is clear that

$$\delta z = f(x+\delta x, y+\delta y) - f(x, y).$$

We first approach the problem geometrically (see Fig. 5.6). Let $P(x, y, z)$ be any point on the surface $z = f(x, y)$, and

$$Q(x+\delta x, y+\delta y, z+\delta z)$$

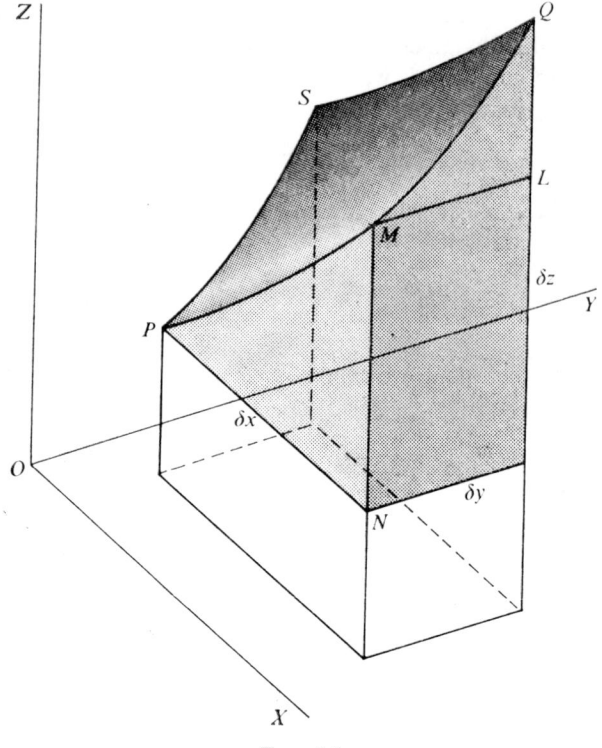

FIG. 5.6

a neighbouring point. From Fig. 5.6, $\delta z = NM + LQ$. Now, since the gradient of the curve PM at P is $\left(\frac{\partial z}{\partial x}\right)_P$, $NM = \left(\frac{\partial z}{\partial x}\right)_P \delta x$ approximately. Since the gradient of the curve MQ at M is $\left(\frac{\partial z}{\partial y}\right)_M$, $LQ = \left(\frac{\partial z}{\partial y}\right)_M \delta y$ approximately and if we assume that the gradient of the curve MQ at M is approximately equal to the gradient of the curve PS at P, we may write $LQ = \left(\frac{\partial z}{\partial y}\right)_P \delta y$ approximately. Hence

$$\delta z = \left(\frac{\partial z}{\partial x}\right)_P \delta x + \left(\frac{\partial z}{\partial y}\right)_P \delta y$$

approximately.

FUNCTIONS OF MORE THAN ONE VARIABLE

We shall now prove the result analytically and make clear the meaning to be attached to the word 'approximately' in the formula. We shall assume that $f(x, y)$ has continuous partial derivatives in the region of the point under consideration. We write δz in the form

$$\delta z = f(x+\delta x, y+\delta y) - f(x, y)$$
$$= \{f(x+\delta x, y+\delta y) - f(x+\delta x, y)\} + \{f(x+\delta x, y) - f(x, y)\}$$

Now

$$\left(\frac{\partial z}{\partial y}\right)_{(x+\delta x, y)} = \lim_{\delta y \to 0} \frac{f(x+\delta x, y+\delta y) - f(x+\delta x, y)}{\delta y},$$

from which it follows that

$$\frac{f(x+\delta x, y+\delta y) - f(x+\delta x, y)}{\delta y} = \left(\frac{\partial z}{\partial y}\right)_{(x+\delta x, y)} + \varepsilon,$$

where $\varepsilon \to 0$ as $\delta y \to 0$. Further, if $\dfrac{\partial z}{\partial y}$ is a continuous function of x and y, then

$$\left(\frac{\partial z}{\partial y}\right)_{(x+\delta x, y)} = \left(\frac{\partial z}{\partial y}\right)_{(x, y)} + \varepsilon_1,$$

where $\varepsilon_1 \to 0$ as $\delta x \to 0$. Thus

$$\frac{f(x+\delta x, y+\delta y) - f(x+\delta x, y)}{\delta y} = \left(\frac{\partial z}{\partial y}\right)_{(x, y)} + \varepsilon_2,$$

where $\varepsilon_2 \to 0$ as δx and $\delta y \to 0$. Hence

$$f(x+\delta x, y+\delta y) - f(x+\delta x, y) = \left(\frac{\partial z}{\partial y}\right)_{(x, y)} \delta y + \varepsilon_2 \, \delta y.$$

We prove in a similar manner that

$$f(x+\delta x, y) - f(x, y) = \left(\frac{\partial z}{\partial x}\right)_{(x, y)} \delta x + \varepsilon_3 \, \delta x,$$

where $\varepsilon_3 \to 0$ as δx and $\delta y \to 0$. Thus

$$\delta z = \frac{\partial z}{\partial x}\delta x + \frac{\partial z}{\partial y}\delta y + \varepsilon_3 \, \delta x + \varepsilon_2 \, \delta y,$$

where ε_3 and ε_2 tend to 0 as δx and δy tend to 0, which gives the precise meaning to be attached to the formula:

$$\delta z = \frac{\partial z}{\partial x}\delta x + \frac{\partial z}{\partial y}\delta y$$

approximately. We sometimes write the result in the form

$$dz = \frac{\partial z}{\partial x}dx + \frac{\partial z}{\partial y}dy,$$

or

$$df = \frac{\partial f}{\partial x}dx + \frac{\partial f}{\partial y}dy.$$

The result may be extended to functions of three or more variables. For example, if $z = f(x, y, u, v)$, where x, y, u, v are independent variables, then

$$\delta z = \frac{\partial z}{\partial x}\delta x + \frac{\partial z}{\partial y}\delta y + \frac{\partial z}{\partial u}\delta u + \frac{\partial z}{\partial v}\delta v.$$

If $z = f(x, y, u, v)$, where u and v are functions of x and y, then we use the alternative form

$$\delta z = \frac{\partial f}{\partial x}\delta x + \frac{\partial f}{\partial y}\delta y + \frac{\partial f}{\partial u}\delta u + \frac{\partial f}{\partial v}\delta v.$$

The results established in this section are the basis of the technique of partial differentiation and have a fundamental importance. The results are also useful in estimating errors arising from calculations based on experimental data.

Example. The well-known result for the time of oscillation of a simple pendulum of length l, $T = 2\pi\sqrt{(l/g)}$, leads to

$$g = \frac{4\pi^2 l}{T^2},$$

and can be used to calculate the value of g from experimental observations of l and T. It is important to know the size of the error in the calculated value of g caused by errors in the measured values of l and T.

If we assume a 1 per cent error in the measured value of l, and a 2 per cent error in the measured value of T, we can find the error in g as follows.

$$\log g = \log(4\pi^2) + \log l - 2\log T.$$

Hence

$$\frac{1}{g}\frac{\partial g}{\partial l} = \frac{1}{l},$$

and

$$\frac{1}{g}\frac{\partial g}{\partial T} = -\frac{2}{T}.$$

FUNCTIONS OF MORE THAN ONE VARIABLE

Hence, on using the result

$$\delta g = \frac{\partial g}{\partial l}\delta l + \frac{\partial g}{\partial T}\delta T,$$

we have

$$\frac{\delta g}{g} = \frac{\delta l}{l} - 2\frac{\delta T}{T}.$$

Now $\frac{\delta l}{l}$ varies between $-0\cdot01$ and $0\cdot01$ and $\frac{\delta T}{T}$ varies between $-0\cdot02$ and $0\cdot02$. Hence $\frac{\delta g}{g}$ varies between $-0\cdot05$ and $0\cdot05$. It follows that the maximum percentage error in the calculated value of g is approximately 5 per cent.

Example. We find from tables that $\sin 16° = 0\cdot2756$ and

$$\cos 16° = 0\cdot9613$$

correct to four decimal places. From these results we can calculate the value of $\tan 16°$ as $\frac{0\cdot2756}{0\cdot9613} = 0\cdot2867$, to four decimal places.

It is of practical importance to have an estimate of the possible error in the value of $\tan 16°$, so calculated, caused by possible errors in the values used for $\sin 16°$ and $\cos 16°$. If we denote $\sin \theta$ by x, $\cos \theta$ by y, and $\tan \theta$ by w, then

$$w = \frac{x}{y},$$

whence

$$\delta w = \frac{1}{y}\delta x - \frac{x}{y^2}\delta y.$$

In this problem the value of δx lies between $-0\cdot000\ 05$ and $0\cdot000\ 05$; the value of δy lies between $-0\cdot000\ 05$ and $0\cdot000\ 05$. Hence the greatest value of δw is

$$\frac{1}{0\cdot9613}(0\cdot000\ 05) - \frac{0\cdot2756}{(0\cdot9613)^2}(-0\cdot000\ 05) = 0\cdot000\ 067.$$

Since the possible error exceeds $0\cdot000\ 05$, the result calculated for $\tan 16°$ cannot be relied upon as correct to four decimal places.

Normals to surfaces

Let $P(x, y, z)$ be any point on the surface $f(x, y, z) = 0$. We shall find the direction cosines of the normal to the surface at P. Let

$Q(x+\delta x, y+\delta y, z+\delta z)$ be a neighbouring point on the surface. Then the direction-cosines of PQ are proportional to $\delta x, \delta y, \delta z$. The line PN with direction-cosines l, m, n will be perpendicular to PQ if

$$l\, \delta x + m\, \delta y + n\, \delta z = 0. \tag{1}$$

Now

$$\frac{\partial f}{\partial x}\delta x + \frac{\partial f}{\partial y}\delta y + \frac{\partial f}{\partial z}\delta z = 0. \tag{2}$$

Hence, equation (1) will be satisfied for *all* sufficiently small values of $\delta x, \delta y$ if

$$\frac{l}{\frac{\partial f}{\partial x}} = \frac{m}{\frac{\partial f}{\partial y}} = \frac{n}{\frac{\partial f}{\partial z}}.$$

It follows that the normal to the surface at P will have direction cosines proportional to

$$\frac{\partial f}{\partial x}, \frac{\partial f}{\partial y}, \frac{\partial f}{\partial z}.$$

The equation of the tangent plane to the surface at (x_1, y_1, z_1) is

$$(x-x_1)\frac{\partial f}{\partial x_1} + (y-y_1)\frac{\partial f}{\partial y_1} + (z-z_1)\frac{\partial f}{\partial z_1} = 0.$$

Exercises 5(f)

1. The area of a triangle ABC is calculated from the formula $\Delta = \frac{1}{2}ab \sin C$, where a, b, C, are measured. Find an expression for the percentage error in Δ in terms of the errors $\delta a, \delta b, \delta C$ in the measured values of a, b, C. If $a = 20$ cm, $b = 30$ cm, each correct to 0·1 cm, and $C = 45°$ correct to 1°, find an estimate for the maximum percentage error in Δ.

2. Find the equation of the tangent plane to the surface $z^5 = xy^3$ at the point (t^2, t, t).

3. A liquid passes through a long narrow tube of length l, radius r. If V is the volume of liquid which escapes per second and p is the pressure difference between the ends of the tube, then it may be proved that μ, the coefficient of viscosity of the liquid, is $\mu = \frac{\pi p r^4}{8lV}$. In an experimental determination of μ, l is measured with 1 per cent accuracy, r and V with 2 per cent accuracy, and p with 3 per cent accuracy. Give an estimate for the maximum percentage error in the calculated value of μ.

4. The formula for the area S of a triangle ABC in terms of a, B, C is

$$S = \tfrac{1}{2}a^2 \frac{\sin B \sin C}{\sin(B+C)}.$$

FUNCTIONS OF MORE THAN ONE VARIABLE

If the side a is kept constant but small changes δB, δC are made in the values of B, C, show that the corresponding small change in S is

$$\delta S = \tfrac{1}{2}a^2 \left\{ \frac{\sin^2 C}{\sin^2 A} \delta B + \frac{\sin^2 B}{\sin^2 A} \delta C \right\} \quad \text{approximately.}$$

The area of a triangular field is calculated from the measured values of $a = 400$ metres, $\angle B = 60°$, $\angle C = 45°$. If the length is correct but the angles are subject to errors of $\tfrac{1}{2}°$, find the calculated value of the area and an approximate value for the maximum error.

5. The angle A of a triangle ABC is calculated from the sides a, b, c. If small changes δa, δb, δc are made in the sides, show that approximately,

$$\delta A = \frac{1}{b \sin C}(\delta a - \cos C . \delta b - \cos B . \delta c).$$

Use this formula to calculate the angles of a triangle in which $a = 4\cdot01$, $b = 4\cdot02$, $c = 3\cdot99$.

6. The top A of a vertical tower AB is observed from two points P, Q on a straight horizontal road through the base B of the tower, P being nearer to B, and $PQ = c$. It is found that $\angle APB = \theta$, $\angle AQB = \phi$. Prove that the height of the tower is $h = \dfrac{c \sin \theta \sin \phi}{\sin(\theta + \phi)}$.

If the measured value of c is correct but the measured values of θ and ϕ are liable to errors of ± 6 minutes, show that the maximum percentage error in the calculated value of the height of the tower is

$$\frac{\pi(\sin^2 \theta + \sin^2 \phi)}{18 \sin \theta \sin \phi \sin(\theta + \phi)}.$$

Technique of partial differentiation

Example. If $z = f(v)$, where v is a given function of x and y, where x and y are independent variables, then z may be considered as a function of the two independent variables x and y. Our aim is to find appropriate formulae for $\partial z/\partial x$ and $\partial z/\partial y$.

The tool in all problems of this type is the fundamental relation between small changes in the variables concerned. Let δx, δy, δv, δz be corresponding small changes in x, y, v, z; δx and δy are independent but δv and δz depend on the values of δx and δy.

Considering z as a function of the single variable v, we have

$$\delta z = f'(v)\delta v + \varepsilon_1 \delta v,$$

where $\varepsilon_1 \to 0$ as $\delta v \to 0$.

Considering v as a function of the two independent variables x and y,

$$\delta v = \frac{\partial v}{\partial x}\delta x + \frac{\partial v}{\partial y}\delta y + \varepsilon_2\,\delta x + \varepsilon_3\,\delta y,$$

where $\varepsilon_2, \varepsilon_3 \to 0$ as $\delta x, \delta y \to 0$. Hence

$$\delta z = f'(v)\frac{\partial v}{\partial x}\delta x + f'(v)\frac{\partial v}{\partial y}\delta y + \varepsilon_1\,\delta v + \varepsilon_2 f'(v)\,\delta x + \varepsilon_3 f'(v)\,\delta y.$$

To find $\dfrac{\partial z}{\partial x}$, we put $\delta y = 0$ in the last equation and obtain

$$\frac{\partial z}{\partial x} = \lim_{\delta x \to 0}\left\{f'(v)\frac{\partial v}{\partial x} + \varepsilon_1\frac{\delta v}{\delta x} + \varepsilon_2 f'(v)\right\}$$

$$= f'(v)\frac{\partial v}{\partial x},$$

since $\dfrac{\delta v}{\delta x}$ and $f'(v)$ are finite. Thus $\dfrac{\partial z}{\partial x} = f'(v)\dfrac{\partial v}{\partial x}$ and similarly

$$\frac{\partial z}{\partial y} = f'(v)\frac{\partial v}{\partial y}.$$

In obtaining results of this kind, we do not normally give the rigorous proof as shown above, unless we are required to do so, but present the argument in the following abbreviated form.

$$\delta z = f'(v)\,\delta v = f'(v)\left\{\frac{\partial v}{\partial x}\delta x + \frac{\partial v}{\partial y}\delta y\right\}$$

$$= \left\{f'(v)\frac{\partial v}{\partial x}\right\}\delta x + \left\{f'(v)\frac{\partial v}{\partial y}\right\}\delta y,$$

whence

$$\frac{\partial z}{\partial x} = f'(v)\frac{\partial v}{\partial x} \quad \text{and} \quad \frac{\partial z}{\partial y} = f'(v)\frac{\partial v}{\partial y}.$$

Applying the results to the case $z = f(x + x^2 y)$, we have

$$\frac{\partial z}{\partial x} = f'(x + x^2 y)(1 + 2xy) \quad \text{and} \quad \frac{\partial z}{\partial y} = f'(x + x^2 y)(x^2).$$

Example. $z = f(x, y, t)$, where t is a function of x and y, x and y being independent variables.

We can regard z as a function of the two independent variables x and y. Let $\delta x, \delta y, \delta t, \delta z$ be corresponding small changes in $x, y,$

FUNCTIONS OF MORE THAN ONE VARIABLE

t, z. Then
$$\delta z = \frac{\partial f}{\partial x}\delta x + \frac{\partial f}{\partial y}\delta y + \frac{\partial f}{\partial t}\delta t.$$

But, since t is a function of x and y,
$$\delta t = \frac{\partial t}{\partial x}\delta x + \frac{\partial t}{\partial y}\delta y.$$

Hence
$$\delta z = \left(\frac{\partial f}{\partial x} + \frac{\partial f}{\partial t}\frac{\partial t}{\partial x}\right)\delta x + \left(\frac{\partial f}{\partial y} + \frac{\partial f}{\partial t}\frac{\partial t}{\partial y}\right)\delta y.$$

It follows that
$$\frac{\partial z}{\partial x} = \frac{\partial f}{\partial x} + \frac{\partial f}{\partial t}\frac{\partial t}{\partial x} \quad \text{and} \quad \frac{\partial z}{\partial y} = \frac{\partial f}{\partial y} + \frac{\partial f}{\partial t}\frac{\partial t}{\partial y}.$$

For example, if $z = xyt^2$, where $t = x^2y$,
$$\frac{\partial z}{\partial x} = yt^2 + 2xyt\frac{\partial t}{\partial x} = yt^2 + 2xyt(2xy) = 5x^4y^3.$$

Alternatively, we can find $\dfrac{\partial z}{\partial x}$ directly by substituting for t in the formula for z, giving $z = x^5y^3$, whence $\dfrac{\partial z}{\partial x} = 5x^4y^3$.

Example. $z = f(u, v)$, where u and v are functions of the two independent variables x and y; then z may be regarded as a function of x and y.

$$\delta z = \frac{\partial f}{\partial u}\delta u + \frac{\partial f}{\partial v}\delta v$$

$$= \frac{\partial f}{\partial u}\left\{\frac{\partial u}{\partial x}\delta x + \frac{\partial u}{\partial y}\delta y\right\} + \frac{\partial f}{\partial v}\left\{\frac{\partial v}{\partial x}\delta x + \frac{\partial v}{\partial y}\delta y\right\}$$

$$= \left\{\frac{\partial f}{\partial u}\frac{\partial u}{\partial x} + \frac{\partial f}{\partial v}\frac{\partial v}{\partial x}\right\}\delta x + \left\{\frac{\partial f}{\partial u}\frac{\partial u}{\partial y} + \frac{\partial f}{\partial v}\frac{\partial v}{\partial y}\right\}\delta y.$$

It follows that
$$\frac{\partial z}{\partial x} = \frac{\partial f}{\partial u}\frac{\partial u}{\partial x} + \frac{\partial f}{\partial v}\frac{\partial v}{\partial x},$$

and
$$\frac{\partial z}{\partial y} = \frac{\partial f}{\partial u}\frac{\partial u}{\partial y} + \frac{\partial f}{\partial v}\frac{\partial v}{\partial y}.$$

For example, if $z = u^2+4v$, where $u = xy+x^2$, $v = x^2y$, then

$$\frac{\partial z}{\partial x} = (2u)(y+2x)+(4)(2xy),$$

and

$$\frac{\partial z}{\partial y} = (2u)(x)+(4)(x^2).$$

Compare these answers with those obtained by substituting for u and v giving z directly as a function of x and y, viz.

$$z = (xy+x^2)^2+4x^2y.$$

Example. The equation $f(x, y, z) = 0$ defines z as a function of the two independent variables x and y. Find $\partial z/\partial x$ and $\partial z/\partial y$. We have

$$\frac{\partial f}{\partial x}\delta x + \frac{\partial f}{\partial y}\delta y + \frac{\partial f}{\partial z}\delta z = 0,$$

which may be expressed in the form

$$\delta z = -\frac{\frac{\partial f}{\partial x}}{\frac{\partial f}{\partial z}}\delta x - \frac{\frac{\partial f}{\partial y}}{\frac{\partial f}{\partial z}}\delta y.$$

It follows that

$$\frac{\partial z}{\partial x} = -\frac{\frac{\partial f}{\partial x}}{\frac{\partial f}{\partial z}} \quad \text{and} \quad \frac{\partial z}{\partial y} = -\frac{\frac{\partial f}{\partial y}}{\frac{\partial f}{\partial z}}.$$

Exercises 5(g)

1. (i) If $z = f(ax+by)$, prove that $b\dfrac{\partial z}{\partial x} = a\dfrac{\partial z}{\partial y}$.

   (ii) If $z = f(e^x y)$, prove that $\dfrac{\partial z}{\partial x} = y\dfrac{\partial z}{\partial y}$.

2. If $z = x^2+yt+xyt^2$, where $t = xy+y^3$, find $\dfrac{\partial z}{\partial x}$ in two ways,

   (i) by using the formula $\dfrac{\partial z}{\partial x} = \dfrac{\partial f}{\partial x}+\dfrac{\partial f}{\partial t}\dfrac{\partial t}{\partial x}$,

   (ii) by expressing z directly as a function of x and y.

   Show that the two answers agree. Repeat for $\dfrac{\partial z}{\partial y}$.

FUNCTIONS OF MORE THAN ONE VARIABLE

3. If $z = f(x, y)$, where x and y are functions of t, prove that
$$\frac{dz}{dt} = \frac{\partial f}{\partial x}\frac{dx}{dt} + \frac{\partial f}{\partial y}\frac{dy}{dt},$$
and find a similar expression for $\frac{d^2z}{dt^2}$.

4. If $z^2 = xyf(x^2-y^2)$, where f is an arbitrary function, and x and y are independent variables, prove that
$$2xy\left(x\frac{\partial z}{\partial y} + y\frac{\partial z}{\partial x}\right) = z(x^2+y^2).$$

5. The equation $f(x, y) = 0$ defines y as a function of x. Prove that
$$\frac{dy}{dx} = -\frac{\partial f}{\partial x}\bigg/\frac{\partial f}{\partial y}.$$

6. If $V = f(x^2+y^2)$ show that $y\frac{\partial V}{\partial x} - x\frac{\partial V}{\partial y} = 0$. Prove also that
$$y^2\frac{\partial^2 V}{\partial x^2} - 2xy\frac{\partial^2 V}{\partial x \partial y} + x^2\frac{\partial^2 V}{\partial y^2} = x\frac{\partial V}{\partial x} + y\frac{\partial V}{\partial y}.$$

7. If $z = f(u, v, w)$, where u, v, w are functions of the two independent variables x and y, find formulae for $\partial z/\partial x$ and $\partial z/\partial y$.

8. Find the partial differential equation satisfied by
$$y^2 + xyz = f(x^2 + yz),$$
where x and y are independent variables, z is the dependent variable, and f is an arbitrary function.

9. If $z = (x+y)\phi\left(\frac{x}{y}\right)$, where ϕ is an arbitrary function, prove that
$$x\frac{\partial z}{\partial x} + y\frac{\partial z}{\partial y} = z \quad \text{and} \quad x^2\frac{\partial^2 z}{\partial x^2} + 2xy\frac{\partial^2 z}{\partial x \partial y} + y^2\frac{\partial^2 z}{\partial y^2} = 0.$$

10. If $z = f(x^2y)$, where f is an arbitrary function, prove that
$$x\frac{\partial z}{\partial x} = 2y\frac{\partial z}{\partial y}.$$

Hence, or otherwise, find the partial differential equation satisfied by $z = f(\log \sin(x^2y) + 3x^4y^2)$.

Find the partial differential equation satisfied by $z = f(x^2y + xy^2)$, where f is an arbitrary function.

11. $z = uv$, where $u = x+y$, $v = x^2+y^2$, x and y being independent variables. Find $\partial z/\partial x$ in two ways:

(i) by using the formula $\dfrac{\partial z}{\partial x} = \dfrac{\partial z}{\partial u}\dfrac{\partial u}{\partial x} + \dfrac{\partial z}{\partial v}\dfrac{\partial v}{\partial x}$,

(ii) by expressing z directly in terms of x and y. Repeat for $\partial z/\partial y$.

12. t is a function of x and y defined by the equation $f(t+x, t+y) = 0$. Prove that $\dfrac{\partial t}{\partial x} + \dfrac{\partial t}{\partial y} = -1$.

Extension of the notation for partial differential coefficients

If we are given two relations of the type $f(x, y, z, u, v) = 0$, $\phi(x, y, z, u, v) = 0$, we can eliminate one of the variables u, v, x, y from the pair of equations and regard z as a function of the three independent variables x, y, u, or x, y, v, or x, u, v, or y, u, v. $\dfrac{\partial z}{\partial x}$ will have a different meaning in the first three cases and no meaning in the fourth case. It is of prime importance, therefore, in a problem like this, to realize clearly which three variables we are regarding z as a function of. We may state this in words or by use of an appropriate notation. We shall denote the three possible interpretations of $\dfrac{\partial z}{\partial x}$ corresponding to the first three cases given above as

$$\left(\dfrac{\partial z}{\partial x}\right)_{y,u}, \left(\dfrac{\partial z}{\partial x}\right)_{y,v}, \left(\dfrac{\partial z}{\partial x}\right)_{u,v}.$$

Example. If $z = x+2y+3u+4v$ and $z = 4x+3y+2u+v$, we may express z in the following forms: $z = \tfrac{1}{3}(15x+10y+5u)$, where x, y, u are independent variables; $z = 10x+5y-5v$, where x, y, v are independent variables; $z = -5x+5u+10v$, where x, u, v are independent variables. It follows that

$$\left(\dfrac{\partial z}{\partial x}\right)_{y,u} = 5, \quad \left(\dfrac{\partial z}{\partial x}\right)_{y,v} = 10, \quad \text{and} \quad \left(\dfrac{\partial z}{\partial x}\right)_{u,v} = -5.$$

Example. If $f(x, y, z)$ is a differentiable function of x, y, z such that $\dfrac{\partial f}{\partial x}, \dfrac{\partial f}{\partial y}, \dfrac{\partial f}{\partial z}$ do not vanish, and if the equation $f(x, y, z) = 0$ can be solved to determine each of the variables x, y, z as a

FUNCTIONS OF MORE THAN ONE VARIABLE

differentiable function of the other two, prove that

$$\left(\frac{\partial z}{\partial y}\right)_x \left(\frac{\partial x}{\partial z}\right)_y \left(\frac{\partial y}{\partial x}\right)_z = -1.$$

The fundamental equation

$$\frac{\partial f}{\partial x}\delta x + \frac{\partial f}{\partial y}\delta y + \frac{\partial f}{\partial z}\delta z = 0$$

can be written in the forms

$$\delta z = -\frac{\frac{\partial f}{\partial x}}{\frac{\partial f}{\partial z}}\delta x - \frac{\frac{\partial f}{\partial y}}{\frac{\partial f}{\partial z}}\delta y,$$

$$\delta x = -\frac{\frac{\partial f}{\partial z}}{\frac{\partial f}{\partial x}}\delta z - \frac{\frac{\partial f}{\partial y}}{\frac{\partial f}{\partial x}}\delta y,$$

$$\delta y = -\frac{\frac{\partial f}{\partial x}}{\frac{\partial f}{\partial y}}\delta x - \frac{\frac{\partial f}{\partial z}}{\frac{\partial f}{\partial y}}\delta z.$$

It follows from these results that

$$\left(\frac{\partial z}{\partial y}\right)_x = -\frac{\frac{\partial f}{\partial y}}{\frac{\partial f}{\partial z}},$$

$$\left(\frac{\partial x}{\partial z}\right)_y = -\frac{\frac{\partial f}{\partial z}}{\frac{\partial f}{\partial x}},$$

and

$$\left(\frac{\partial y}{\partial x}\right)_z = -\frac{\frac{\partial f}{\partial x}}{\frac{\partial f}{\partial y}}.$$

Hence $\left(\dfrac{\partial z}{\partial y}\right)_x \left(\dfrac{\partial x}{\partial z}\right)_y \left(\dfrac{\partial y}{\partial x}\right)_z = -1.$

Example. $x = f(u, v)$, $y = \phi(u, v)$, where f and ϕ are differentiable functions of u and v. If the equations are such that u and v can be expressed as single-valued differentiable functions of x and y find $\dfrac{\partial u}{\partial x}$ and $\dfrac{\partial u}{\partial y}$ in terms of the partial derivatives of f and ϕ with respect to u and v.

If δx, δy, δu, δv are corresponding small increments in x, y, u, v we have

$$\delta x = \frac{\partial f}{\partial u}\delta u + \frac{\partial f}{\partial v}\delta v,$$

$$\delta y = \frac{\partial \phi}{\partial u}\delta u + \frac{\partial \phi}{\partial v}\delta v.$$

It follows that, on solving these equations for δu and δv,

$$\delta u = \frac{\dfrac{\partial \phi}{\partial v}\delta x - \dfrac{\partial f}{\partial v}\delta y}{\dfrac{\partial f}{\partial u}\dfrac{\partial \phi}{\partial v} - \dfrac{\partial f}{\partial v}\dfrac{\partial \phi}{\partial u}},$$

provided the denominator is not zero. It follows that

$$\frac{\partial u}{\partial x} = \frac{\dfrac{\partial \phi}{\partial v}}{\dfrac{\partial f}{\partial u}\dfrac{\partial \phi}{\partial v} - \dfrac{\partial f}{\partial v}\dfrac{\partial \phi}{\partial u}},$$

and

$$\frac{\partial u}{\partial y} = \frac{-\dfrac{\partial f}{\partial v}}{\dfrac{\partial f}{\partial u}\dfrac{\partial \phi}{\partial v} - \dfrac{\partial f}{\partial v}\dfrac{\partial \phi}{\partial u}},$$

provided $\dfrac{\partial f}{\partial u}\dfrac{\partial \phi}{\partial v} - \dfrac{\partial f}{\partial v}\dfrac{\partial \phi}{\partial u} \neq 0.$

Obtain similar expressions for $\dfrac{\partial v}{\partial x}$, $\dfrac{\partial v}{\partial y}$.

Laplace's equation

Laplace (1749–1827) found that the same equation occurred frequently in investigations of a wide field of problems including the

FUNCTIONS OF MORE THAN ONE VARIABLE 149

theory of gravitational potential, electromagnetic theory, and hydrodynamics. That equation, which is associated with his name, is

$$\frac{\partial^2 V}{\partial x^2}+\frac{\partial^2 V}{\partial y^2}+\frac{\partial^2 V}{\partial z^2}=0,$$

where V is a function of the three independent variables x, y, z.

We shall consider the two-dimensional form of Laplace's equation,

$$\frac{\partial^2 V}{\partial x^2}+\frac{\partial^2 V}{\partial y^2}=0,$$

where V is a function of x and y, x and y being Cartesian coordinates. In applying this equation to some problems, e.g. the motion of a liquid past a stationary sphere, it is helpful to write the equation in terms of polar coordinates r, θ, instead of Cartesian coordinates x, y. It is interesting to note that Laplace's first statement of his equation was in polar form.

Transformation of $\dfrac{\partial^2 V}{\partial x^2}+\dfrac{\partial^2 V}{\partial y^2}$ *to polar form*

On applying the result $\dfrac{\partial V}{\partial x}=\dfrac{\partial V}{\partial r}\dfrac{\partial r}{\partial x}+\dfrac{\partial V}{\partial \theta}\dfrac{\partial \theta}{\partial x}$, it is necessary to recognize that we must express r and θ as functions of x and y, namely, $r^2 = x^2+y^2$, $\theta = \tan^{-1}\left(\dfrac{y}{x}\right)$. Thus $2r\dfrac{\partial r}{\partial x} = 2x$, whence $\dfrac{\partial r}{\partial x}=\dfrac{x}{r}$, and

$$\frac{\partial \theta}{\partial x} = \frac{1}{1+\left(\dfrac{y}{x}\right)^2}\left(-\frac{y}{x^2}\right) = -\frac{y}{r^2}.$$

Hence

$$\frac{\partial V}{\partial x}=\frac{\partial V}{\partial r}\frac{x}{r}-\frac{\partial V}{\partial \theta}\frac{y}{r^2}.$$

On differentiating this equation partially with respect to x,

$$\frac{\partial^2 V}{\partial x^2} = \frac{\partial V}{\partial r}\left(\frac{1}{r}-\frac{x}{r^2}\frac{x}{r}\right)+\frac{x}{r}\left\{\frac{\partial^2 V}{\partial r^2}\frac{x}{r}+\frac{\partial^2 V}{\partial r \partial \theta}\left(-\frac{y}{r^2}\right)\right\}+$$

$$+\frac{\partial V}{\partial \theta}\frac{2y}{r^3}\frac{x}{r}-\frac{y}{r^2}\left\{\frac{\partial^2 V}{\partial r \partial \theta}\frac{x}{r}+\frac{\partial^2 V}{\partial \theta^2}\left(-\frac{y}{r^2}\right)\right\}$$

$$=\frac{\partial V}{\partial r}\left(\frac{y^2}{r^3}\right)+\frac{\partial V}{\partial \theta}\left(\frac{2xy}{r^4}\right)+\frac{x^2}{r^2}\frac{\partial^2 V}{\partial r^2}-\frac{2xy}{r^3}\frac{\partial^2 V}{\partial r \partial \theta}+\frac{y^2}{r^4}\frac{\partial^2 V}{\partial \theta^2}.$$

We prove in a similar manner that

$$\frac{\partial^2 V}{\partial y^2} = \frac{\partial V}{\partial r}\left(\frac{x^2}{r^3}\right) - \frac{\partial V}{\partial \theta}\left(\frac{2xy}{r^4}\right) + \frac{y^2}{r^2}\frac{\partial^2 V}{\partial r^2} + \frac{2xy}{r^3}\frac{\partial^2 V}{\partial r \partial \theta} + \frac{x^2}{r^4}\frac{\partial^2 V}{\partial \theta^2}.$$

Hence

$$\frac{\partial^2 V}{\partial x^2} + \frac{\partial^2 V}{\partial y^2} = \frac{1}{r}\frac{\partial V}{\partial r} + \frac{\partial^2 V}{\partial r^2} + \frac{1}{r^2}\frac{\partial^2 V}{\partial \theta^2}.$$

It is somewhat simpler to transform the r, θ form of Laplace's equation into the Cartesian form as follows.

On applying the result $\dfrac{\partial V}{\partial r} = \dfrac{\partial V}{\partial x}\dfrac{\partial x}{\partial r} + \dfrac{\partial V}{\partial y}\dfrac{\partial y}{\partial r}$, we must express x and y as functions of r and θ, viz. $x = r\cos\theta$, $y = r\sin\theta$. Then

$$\frac{\partial V}{\partial r} = \frac{\partial V}{\partial x}(\cos\theta) + \frac{\partial V}{\partial y}(\sin\theta),$$

$$\frac{\partial^2 V}{\partial r^2} = \cos\theta\left\{\frac{\partial^2 V}{\partial x^2}\cos\theta + \frac{\partial^2 V}{\partial x \partial y}\sin\theta\right\} + \sin\theta\left\{\frac{\partial^2 V}{\partial x \partial y}\cos\theta + \frac{\partial^2 V}{\partial y^2}\sin\theta\right\}.$$

$$\frac{\partial V}{\partial \theta} = \frac{\partial V}{\partial x}\frac{\partial x}{\partial \theta} + \frac{\partial V}{\partial y}\frac{\partial y}{\partial \theta} = \frac{\partial V}{\partial x}(-r\sin\theta) + \frac{\partial V}{\partial y}(r\cos\theta).$$

$$\frac{\partial^2 V}{\partial \theta^2} = -r\frac{\partial V}{\partial x}\cos\theta - r\sin\theta\left\{\frac{\partial^2 V}{\partial x^2}(-r\sin\theta) + \frac{\partial^2 V}{\partial x \partial y}(r\cos\theta)\right\} +$$

$$+ r\frac{\partial V}{\partial y}\sin\theta + r\cos\theta\left\{\frac{\partial^2 V}{\partial x \partial y}(-r\sin\theta) + \frac{\partial^2 V}{\partial y^2}(r\cos\theta)\right\}.$$

On combining these results, we can show that

$$\frac{1}{r}\frac{\partial V}{\partial r} + \frac{\partial^2 V}{\partial r^2} + \frac{1}{r^2}\frac{\partial^2 V}{\partial \theta^2} = \frac{\partial^2 V}{\partial x^2} + \frac{\partial^2 V}{\partial y^2}.$$

If $z = f(u, v)$ and we are given a pair of relations between u, v, x, y, then we may regard z as a function of x and y; but in applying the formula

$$\frac{\partial z}{\partial x} = \frac{\partial f}{\partial u}\frac{\partial u}{\partial x} + \frac{\partial f}{\partial v}\frac{\partial v}{\partial x},$$

it is desirable to have u and v explicitly in terms of x and y, as the two-way transformation of Laplace's equation illustrates. We could, of course, find $\partial r/\partial x$ from $x = r\cos\theta$, $y = r\sin\theta$ by using

FUNCTIONS OF MORE THAN ONE VARIABLE 151

the results proved on p. 148, which give

$$\frac{\partial r}{\partial x} = \frac{\dfrac{\partial y}{\partial \theta}}{\dfrac{\partial x}{\partial r}\dfrac{\partial y}{\partial \theta} - \dfrac{\partial x}{\partial \theta}\dfrac{\partial y}{\partial r}} = \frac{r\cos\theta}{(\cos\theta)(r\cos\theta)-(-r\sin\theta)(\sin\theta)}$$

$$= \cos\theta = \frac{x}{r}.$$

But it is obviously more convenient to express r in terms of x and y.

Exercises 5(h)

1. Prove in two ways that the differential equation $\dfrac{\partial^2 V}{\partial x^2}+\dfrac{\partial^2 V}{\partial y^2}=0$ is satisfied by $V = \dfrac{xy}{(x^2+y^2)^2}$, (i) directly (ii) by converting the differential equation and the proposed solution into polar form.

2. Regarding z as a function of the two independent variables x and y, find the first order partial differential equation satisfied by z subject to the following relations, f denoting an arbitrary function:

(i) $y^2 - xz = f(x^2 - yz)$,
(ii) $z + xy = f(xyz)$,
(iii) $x^2 + y^2 + z^2 = f(xy + yz + zx)$.

3. Find the equation of the tangent plane at the point (a, a, a) on the surface $x^4 + y^4 + z^4 = 3a^4$.

4. If $u = f(r)$ and $x = r\cos\theta$, $y = r\sin\theta$, prove that

$$\frac{\partial^2 u}{\partial x^2}+\frac{\partial^2 u}{\partial y^2}=f''(r)+\frac{1}{r}f'(r).$$

5. If $V = f\left(xz, \dfrac{y}{z}\right)$, where x, y, z are independent variables, prove that

$$z\frac{\partial V}{\partial z} = x\frac{\partial V}{\partial x} - y\frac{\partial V}{\partial y}.$$

6. If $x = \tfrac{1}{2}\log(u^2+v^2)$, $y = \tan^{-1}\left(\dfrac{v}{u}\right)$, use the results established on page 148 to find expressions for $\dfrac{\partial u}{\partial x}, \dfrac{\partial u}{\partial y}, \dfrac{\partial v}{\partial x}, \dfrac{\partial v}{\partial y}$. Compare your answers with those obtained by finding explicit formulae for u and v in terms of x and y, and then finding the partial derivatives directly.

7. If $z = f(xy) + \phi(x, y)$, where f and ϕ are arbitrary functions, show that $w \equiv x\dfrac{\partial z}{\partial x} - y\dfrac{\partial z}{\partial y}$ is independent of the choice of f. Find w when $\phi(x, y) = xy^2 e^{xy}$.

8. If $w = x^n f(u, v)$, where $u = \dfrac{y}{x}$, $v = \dfrac{z}{x}$ and x, y, z are independent variables, prove that
$$x\frac{\partial w}{\partial x} + y\frac{\partial w}{\partial y} + z\frac{\partial w}{\partial z} = nw.$$

9. If $V = f(x^2 + y^2 + z^2)$, where x, y, z are independent variables, show that
$$\frac{\partial^2 V}{\partial x^2} + \frac{\partial^2 V}{\partial y^2} + \frac{\partial^2 V}{\partial z^2} = 4(x^2+y^2+z^2)f''(x^2+y^2+z^2) + 6f'(x^2+y^2+z^2).$$

10. If $z = f(x, y, u, v)$, where u and v are functions of x and y, prove that
$$\frac{\partial z}{\partial x} = \frac{\partial f}{\partial x} + \frac{\partial f}{\partial u}\frac{\partial u}{\partial x} + \frac{\partial f}{\partial v}\frac{\partial v}{\partial x},$$
and write down a corresponding formula for $\dfrac{\partial z}{\partial y}$.

Illustrate the result by finding $\dfrac{\partial z}{\partial x}$ (i) by use of the formula, (ii) directly by expressing z in terms of x and y only, in the case when $z = xyuv$, $u = x^2 + y^2$, $v = xy^2$.

11. If $v = f(x^2 - y^2)$, where f is any function, prove that
$$y^2\frac{\partial^2 v}{\partial x^2} + 2xy\frac{\partial^2 v}{\partial x \partial y} + x^2\frac{\partial^2 v}{\partial y^2} + x\frac{\partial v}{\partial x} + y\frac{\partial v}{\partial y} = 0.$$

12. u and v are functions of the two independent variables x and y. If these equations can be solved to give x and y as single-valued differentiable functions of u and v, show that
$$\frac{\partial u}{\partial x}\frac{\partial x}{\partial u} + \frac{\partial v}{\partial x}\frac{\partial x}{\partial v} = 1 \quad \text{and} \quad \frac{\partial u}{\partial x}\frac{\partial y}{\partial u} + \frac{\partial v}{\partial x}\frac{\partial y}{\partial v} = 0.$$

13. If $v = f(x, y, z)$, where $g(x, y, z) = 0$ and $h(x, y, z) = 0$, and we assume that the last two equations can be solved to give y and z as functions of x, then we can express v as a function of the single variable x. We assume that all the functions concerned are differentiable. Find an expression for dv/dx in terms of the partial differential coefficients of the functions f, g, h.

14. If $f(x, y, z) = 0$, prove that $\left(\dfrac{\partial z}{\partial x}\right)_y = 1 \bigg/ \left(\dfrac{\partial x}{\partial z}\right)_y$.

Miscellaneous problems on Chapter 5

1. Find the envelope of the family of straight lines
$$y = tx - t^4,$$

FUNCTIONS OF MORE THAN ONE VARIABLE

where t is a variable parameter. Sketch some of the lines of the system and the envelope.

2. $V = f(x, y, z)$, where $z = xy^2$. Prove that
$$\frac{\partial V}{\partial x} - \frac{\partial f}{\partial x} = y^2 \frac{\partial f}{\partial z},$$
and find a similar expression for $\dfrac{\partial V}{\partial y} - \dfrac{\partial f}{\partial y}$.

3. Find the equation of the tangent plane at the point (t, t^2, t^5) on the surface $z = x^3 y$. Consider the lines of intersection of the tangent planes with the plane $z = 0$ and prove that the envelope of the family of lines is the parabola,
$$4y = -3x^2, \quad z = 0.$$

4. If $v = y^m f\left(\dfrac{x}{y}\right)$, prove that $x\dfrac{\partial v}{\partial x} + y\dfrac{\partial v}{\partial y} = mv$. If $\dfrac{\partial^2 v}{\partial x \partial y} = 0$, and t denotes $\dfrac{x}{y}$, show that $(m-1)\dfrac{df}{dt} = t\dfrac{d^2 f}{dt^2}$, and hence verify that v is of the form $ax^m + by^m$, where a and b are constants.

5. Show that the equation $\dfrac{\partial^2 z}{\partial t^2} = k^2 \dfrac{\partial^2 z}{\partial x^2}$ is satisfied by
$$z = f(x+kt) + \phi(x-kt),$$
where f and ϕ are arbitrary functions. Obtain a solution of the differential equation which satisfies $z = e^x - 1$ and $\dfrac{\partial z}{\partial t} = x$ when $t = 0$.

6. If $w = 2uv + f\left(\dfrac{u}{v}\right)$, where u and v are independent variables and f is an arbitrary function, prove that
$$u^2 \frac{\partial^2 w}{\partial u^2} + 2uv \frac{\partial^2 w}{\partial u \partial v} + v^2 \frac{\partial^2 w}{\partial v^2} = 4uv.$$

7. u and v are two variables, given in terms of the independent variables x and y, by the laws
$$u = x^2 - y^2, \quad v = 2xy.$$
A function of x and y, $f(x, y)$, is transformed using these results into a function of u and v, $\phi(u, v)$, so that we may write $f(x, y) \equiv \phi(u, v)$. Prove that

(i) $\left(\dfrac{\partial f}{\partial x}\right)^2 + \left(\dfrac{\partial f}{\partial y}\right)^2 = 4(u^2+v^2)^{\frac{1}{2}} \left\{ \left(\dfrac{\partial \phi}{\partial u}\right)^2 + \left(\dfrac{\partial \phi}{\partial v}\right)^2 \right\},$

(ii) $\dfrac{\partial^2 f}{\partial x^2} + \dfrac{\partial^2 f}{\partial y^2} = 4(u^2+v^2)^{\frac{1}{2}} \left\{ \dfrac{\partial^2 \phi}{\partial u^2} + \dfrac{\partial^2 \phi}{\partial v^2} \right\}.$

8. A current flows in a circular wire centre O, radius r. The resulting magnetic intensity I at a point P on a line through O perpendicular to the plane of the wire, where $OP = x$, is given by
$$I = \frac{kr^2}{(r^2+x^2)^{\frac{3}{2}}},$$
where k is a constant. I is calculated from the measured values $r = 10$ cm, $x = 20$ cm, both measurements being correct to within ± 1 mm. Find an approximate value of the greatest possible percentage error in I.

9. If r and t are independent variables, and
$$v = \frac{1}{r}f(ct+r) + \frac{1}{r}\phi(ct-r),$$
prove that $\dfrac{\partial^2 v}{\partial t^2} - c^2 \dfrac{\partial^2 v}{\partial r^2} - \dfrac{2c^2}{r}\dfrac{\partial v}{\partial r} = 0$.

10. If $f(u)$ is a polynomial in u and $z^2 = f(u)f(v)$, where u and v are independent variables, prove that
$$z\frac{\partial}{\partial u}\left\{\frac{f(u)}{z(u-v)}\right\} - z\frac{\partial}{\partial v}\left\{\frac{f(v)}{z(v-u)}\right\}$$
is a polynomial in u and v.

11. θ and ϕ are independent variables and
$$z = \cos\theta\cos\alpha + \sin\theta\sin\alpha\cos(\phi-\beta),$$
α and β being constants. Prove that
$$\left(\frac{\partial z}{\partial \theta}\right)^2 + \operatorname{cosec}^2\theta\left(\frac{\partial z}{\partial \phi}\right)^2 = 1 - z^2,$$
and
$$\frac{\partial^2 z}{\partial \theta^2} + \cot\theta\frac{\partial z}{\partial \theta} + \operatorname{cosec}^2\theta\frac{\partial^2 z}{\partial \phi^2} = -2z.$$

12. Find the stationary points of the surface
$$z = (x+2y-2)(x-y)(x+y).$$
Discover whether the points are maximum points, minimum points, or saddle points in two ways:
 (i) by investigating the behavior of z in the neighbourhood of the points;
 (ii) by considering the sign of z in the regions into which the XOY plane is divided by the lines
$$x+2y-2 = 0; \qquad x-y = 0; \qquad x+y = 0.$$

13. Given that u is a function of x and y, prove that
$$\frac{\partial}{\partial y}\left\{f(u)\frac{\partial u}{\partial x}\right\} = \frac{\partial}{\partial x}\left\{f(u)\frac{\partial u}{\partial y}\right\}.$$

FUNCTIONS OF MORE THAN ONE VARIABLE 155

14. If $\dfrac{x^2}{a^2+z}+\dfrac{y^2}{b^2+z}=1$, where x and y are independent variables and z is the dependent variable, prove that

$$\left(\frac{\partial z}{\partial x}\right)^2+\left(\frac{\partial z}{\partial y}\right)^2=2\left(x\frac{\partial z}{\partial x}+y\frac{\partial z}{\partial y}\right).$$

15. $z=f(x,y)$, where $\phi(x,y)=0$. Prove that if z is expressed as a function of the single variable x,

$$\frac{dz}{dx}=\frac{\dfrac{\partial f}{\partial x}\dfrac{\partial \phi}{\partial y}-\dfrac{\partial f}{\partial y}\dfrac{\partial \phi}{\partial x}}{\dfrac{\partial \phi}{\partial y}}.$$

16. Prove in two ways that the equation

$$\frac{\partial^2 V}{\partial x^2}+\frac{\partial^2 V}{\partial y^2}=0$$

is satisfied by

$$V=\left(Ar+\frac{B}{r}\right)\cos\theta,$$

where A and B are arbitrary constants and $x=r\cos\theta$, $y=r\sin\theta$, (i) by converting the suggested solution into Cartesian form, (ii) by converting the differential equation into polar form.

Obtain a solution which vanishes when $r=a$, and tends to x as r tends to infinity.

17. The pair of equations $f(x,y,z)=$ constant, $xyz=$ constant, define y as a function of x. Show that

$$\frac{dy}{dx}=\frac{-y\left(x\dfrac{\partial f}{\partial x}-z\dfrac{\partial f}{\partial z}\right)}{x\left(y\dfrac{\partial f}{\partial y}-z\dfrac{\partial f}{\partial z}\right)}.$$

Check the result by considering the particular case $xy^2z^3=$ constant, $xyz=$ constant.

18. In a triangulation survey the sides b, c of a triangle ABC are measured as 200 metres, 240 metres respectively, and the angle A as 45°. If b and c are both measured too large by 40 cm, what compensating error in the measurement of angle A will lead to a correct value of the side a? What compensating error in the measurement of angle A will lead to a correct value of the area of the triangle?

156   MATHEMATICAL ANALYSIS AND TECHNIQUES

19. If z is a function of x and y and $y = tx$, prove that

$$\left(\frac{\partial z}{\partial x}\right)_{y\text{ const.}} = \left(\frac{\partial z}{\partial x}\right)_{t\text{ const.}} - \frac{t}{x}\left(\frac{\partial z}{\partial t}\right)_{x\text{ const.}}.$$

20. If $z = x^n\{f(x+y) + \phi(y-x)\}$, where f and ϕ are arbitrary functions, prove that

$$\frac{\partial^2 z}{\partial x^2} - \frac{\partial^2 z}{\partial y^2} - \frac{2n}{x}\frac{\partial z}{\partial x} + \frac{n(n+1)}{x^2}z = 0.$$

21. x, y, z are independent variables, t is a function of x, y, z defined by the relation
$$f(x^2+t^2, y^2+t^2, z^2+t^2) = 0.$$
Prove that
$$\frac{1}{t} + \frac{1}{x}\frac{\partial t}{\partial x} + \frac{1}{y}\frac{\partial t}{\partial y} + \frac{1}{z}\frac{\partial t}{\partial z} = 0.$$

22. If $v = f(x-y, y-z, z-x)$, where x, y, z are independent variables, prove that
$$\frac{\partial v}{\partial x} + \frac{\partial v}{\partial y} + \frac{\partial v}{\partial z} = 0.$$

Illustrate by evaluating $\dfrac{\partial v}{\partial x} + \dfrac{\partial v}{\partial y} + \dfrac{\partial v}{\partial z}$ directly in the case

$$v = (x-y)(y-z)e^{z-x}.$$

23. $f(x, y)$ is transformed by the substitutions $x = u^2 + v^2$, $y = 2uv$, into a function of u and v, $\phi(u, v)$, so that $\phi(u, v) \equiv f(x, y)$.
Show that
$$u\frac{\partial \phi}{\partial u} - v\frac{\partial \phi}{\partial v} = 2(u^2 - v^2)\frac{\partial f}{\partial x},$$
and
$$v\frac{\partial \phi}{\partial u} - u\frac{\partial \phi}{\partial v} = 2(v^2 - u^2)\frac{\partial f}{\partial y}.$$

6
Maximum and Minimum Values of a Function of Two Variables

Further topics in partial differentiation

The stationary values of $z = f(x, y)$ are found by solving the equations $\dfrac{\partial z}{\partial x} = 0$ and $\dfrac{\partial z}{\partial y} = 0$, and we have shown how we can determine whether any particular solution of these equations gives a maximum value of z, a minimum value of z, or a saddle-point by considering the behaviour of z in the neighbourhood of the stationary point. We shall now produce a set of simple conditions for diagnosing what sort of stationary point we have. The method depends on the extension of Taylor's theorem to a function of two variables.

Taylor's theorem for a function of two variables

Consider $z = f(x, y)$, where $x = a+ht$, $y = b+kt$, and a, b, h, k are constants. Thus z may be expressed as a function of the single variable t, say $z = \phi(t)$. Applying Maclaurin's theorem to this function, we have, under suitable conditions,

$$z = \phi(0) + t\phi'(0) + \frac{t^2}{2!}\phi''(0) + \cdots.$$

The next step is to express $\phi(0)$, $\phi'(0)$, etc. in terms of $f(x, y)$ and its partial derivatives with respect to x and y.

$$\phi(0) = f(a, b)$$

$$\phi'(0) = \left(\frac{dz}{dt}\right)_{t=0}.$$

Now

$$\frac{dz}{dt} = \frac{\partial f}{\partial x} \cdot \frac{dx}{dt} + \frac{\partial f}{\partial y} \cdot \frac{dy}{dt} = h\frac{\partial f}{\partial x} + k\frac{\partial f}{\partial y}.$$

Hence
$$\phi'(0) = \left(h\frac{\partial f}{\partial x} + k\frac{\partial f}{\partial y}\right)_{x=a, y=b}.$$

$$\frac{d^2 z}{dt^2} = h\left\{\frac{\partial^2 f}{\partial x^2}\cdot\frac{dx}{dt} + \frac{\partial^2 f}{\partial x\,\partial y}\cdot\frac{dy}{dt}\right\} + k\left\{\frac{\partial^2 f}{\partial x\,\partial y}\cdot\frac{dx}{dt} + \frac{\partial^2 f}{\partial y^2}\cdot\frac{dy}{dt}\right\}$$

$$= h^2\frac{\partial^2 f}{\partial x^2} + 2hk\frac{\partial^2 f}{\partial x\,\partial y} + k^2\frac{\partial^2 f}{\partial y^2}.$$

Hence
$$\phi''(0) = \left(h^2\frac{\partial^2 f}{\partial x^2} + 2hk\frac{\partial^2 f}{\partial x\,\partial y} + k^2\frac{\partial^2 f}{\partial y^2}\right)_{x=a, y=b}.$$

In a similar manner we prove that

$$\phi^{(3)}(0) = \left(h^3\frac{\partial^3 f}{\partial x^3} + 3h^2 k\frac{\partial^3 f}{\partial x^2\,\partial y} + 3hk^2\frac{\partial^3 f}{\partial x\,\partial y^2} + k^3\frac{\partial^3 f}{\partial y^3}\right)_{x=a, y=b},$$

and so on. Substituting these values of $\phi(0)$, $\phi'(0)$, etc., in the Maclaurin expansion above, we have

$$f(a+ht, b+kt) = f(a, b) + t\left(h\frac{\partial f}{\partial x} + k\frac{\partial f}{\partial y}\right)_{x=a, y=b} +$$
$$+ \frac{t^2}{2!}\left(h^2\frac{\partial^2 f}{\partial x^2} + 2hk\frac{\partial^2 f}{\partial x\,\partial y} + k^2\frac{\partial^2 f}{\partial y^2}\right)_{x=a, y=b} + \cdots.$$

Putting $t = 1$, we have

$$f(a+h, b+k) = f(a, b) + \left(h\frac{\partial f}{\partial x} + k\frac{\partial f}{\partial y}\right) +$$
$$+ \frac{1}{2!}\left(h^2\frac{\partial^2 f}{\partial x^2} + 2hk\frac{\partial^2 f}{\partial x\,\partial y} + k^2\frac{\partial^2 f}{\partial y^2}\right) + \cdots,$$

where the differential coefficients are evaluated at $x = a$, $y = b$.

We are now in a position to prove sufficient conditions for maximum and minimum values of $z = f(x, y)$. As we have already seen, necessary conditions for maximum and minimum values are $\partial f/\partial x = 0$, $\partial f/\partial y = 0$. If $x = a$, $y = b$ is a solution of these equations, then, by Taylor's theorem,

$$(a+h, b+k) - f(a, b) = \frac{1}{2!}\left(h^2\frac{\partial^2 f}{\partial x^2} + 2hk\frac{\partial^2 f}{\partial x\,\partial y} + k^2\frac{\partial^2 f}{\partial y^2}\right)$$

approximately† for small values of h and k, where the differential coefficients are evaluated at $x = a$, $y = b$. Hence, for all small h and k, the sign of $f(a+h, b+k) - f(a, b)$ is the sign of

$$\lambda^2 \frac{\partial^2 f}{\partial x^2} + 2\lambda \frac{\partial^2 f}{\partial x \, \partial y} + \frac{\partial^2 f}{\partial y^2},$$

where $\lambda = \dfrac{h}{k}$, $(k \neq 0)$.

Now $p\lambda^2 + 2q\lambda + r$ will have a fixed sign for all values of λ if $q^2 < pr$, and the sign will be positive if $p > 0$ and negative if $p < 0$.‡ It follows that $f(a+h, b+k) - f(a, b)$ will have a fixed positive sign for all small values of h and k if

$$\left(\frac{\partial^2 f}{\partial x \, \partial y}\right)^2 < \left(\frac{\partial^2 f}{\partial x^2}\right)\left(\frac{\partial^2 f}{\partial y^2}\right) \quad \text{and} \quad \frac{\partial^2 f}{\partial x^2} > 0,$$

and in this case $x = a$, $y = b$ will clearly give a minimum value of $f(x, y)$. $f(a+h, b+k) - f(a, b)$ will have a fixed negative sign for all small values of h and k if

$$\left(\frac{\partial^2 f}{\partial x \, \partial y}\right)^2 < \left(\frac{\partial^2 f}{\partial x^2}\right)\left(\frac{\partial^2 f}{\partial y^2}\right) \quad \text{and} \quad \frac{\partial^2 f}{\partial x^2} < 0,$$

and in this case $x = a$, $y = b$ will clearly give a maximum value of $f(x, y)$.

It is clear that if

$$\left(\frac{\partial^2 f}{\partial x \, \partial y}\right)^2 > \left(\frac{\partial^2 f}{\partial x^2}\right)\left(\frac{\partial^2 f}{\partial y^2}\right),$$

the sign of $f(a+h, b+k) - f(a, b)$ is not fixed for all small values of h and k and hence the point P on the surface given by $x = a$, $y = b$ gives neither a maximum nor a minimum value of z. Some sections of the surface through P have a minimum at P, other sections have a maximum at P, and other sections have inflexions at P. P is called a saddle-point of the surface.

If

$$\left(\frac{\partial^2 f}{\partial x \, \partial y}\right)^2 = \left(\frac{\partial^2 f}{\partial x^2}\right)\left(\frac{\partial^2 f}{\partial y^2}\right),$$

† This assumes that $\dfrac{\partial^2 f}{\partial x^2}$, $\dfrac{\partial^2 f}{\partial x \, \partial y}$ and $\dfrac{\partial^2 f}{\partial y^2}$ do not *all* vanish when $x = a$, $y = b$, in which case we must take the next term of the Taylor expansion.

‡ This is clear from the fact that

$$p\lambda^2 + 2q\lambda + r \equiv p\left\{\left(\lambda + \frac{q}{p}\right)^2 + \frac{pr - q^2}{p^2}\right\}.$$

the test provides no information. The point in question may be a maximum, a minimum or a saddle-point.

Example. $z = xy(x+y-1)$,

$$\frac{\partial z}{\partial x} = 2xy+y^2-y, \qquad \frac{\partial z}{\partial y} = 2xy+x^2-x,$$

$$\frac{\partial^2 z}{\partial x^2} = 2y, \qquad \frac{\partial^2 z}{\partial x\,\partial y} = 2x+2y-1, \qquad \frac{\partial^2 z}{\partial y^2} = 2x.$$

The solutions of the equations $\frac{\partial z}{\partial x} = 0, \frac{\partial z}{\partial y} = 0$, are $x = 0, y = 0$, and $x = \frac{1}{3}, y = \frac{1}{3}$. Now

Since
$$\left[\frac{\partial^2 z}{\partial x^2}\right]_{0,0} = 0, \quad \left[\frac{\partial^2 z}{\partial x\,\partial y}\right]_{0,0} = -1, \quad \left[\frac{\partial^2 z}{\partial y^2}\right]_{0,0} = 0.$$

$$\left(\frac{\partial^2 z}{\partial x\,\partial y}\right)^2 > \left(\frac{\partial^2 z}{\partial x^2}\right)\left(\frac{\partial^2 z}{\partial y^2}\right),$$

there is a saddle-point at (0, 0, 0). And in the second case

$$\left[\frac{\partial^2 z}{\partial x^2}\right]_{\frac{1}{3},\frac{1}{3}} = \frac{2}{3}, \quad \left[\frac{\partial^2 z}{\partial x\,\partial y}\right]_{\frac{1}{3},\frac{1}{3}} = \frac{1}{3}, \quad \left[\frac{\partial^2 z}{\partial y^2}\right]_{\frac{1}{3},\frac{1}{3}} = \frac{2}{3}.$$

In this case
$$\left(\frac{\partial^2 z}{\partial x\,\partial y}\right)^2 < \left(\frac{\partial^2 z}{\partial x^2}\right)\left(\frac{\partial^2 z}{\partial y^2}\right) \quad \text{and} \quad \frac{\partial^2 z}{\partial x^2} > 0,$$

so it follows that $x = \frac{1}{3}, y = \frac{1}{3}$ gives a minimum value of z.

Example. Find the largest and smallest values of z for the portion of the surface
$$z = y^2(x^2+y^2-4)+4x^2,$$
which lies within the cylinder $x^2+y^2 = 4$.

$$\frac{\partial z}{\partial x} = 2x(y^2+4), \qquad \frac{\partial z}{\partial y} = 2y(x^2+2y^2-4),$$

$$\frac{\partial^2 z}{\partial x^2} = 2y^2+8, \qquad \frac{\partial^2 z}{\partial x\,\partial y} = 4xy, \qquad \frac{\partial^2 z}{\partial y^2} = 2x^2+12y^2-8.$$

The solutions of $\frac{\partial z}{\partial x} = 0, \frac{\partial z}{\partial y} = 0$ are

$$x = 0, y = 0; \ x = 0, y = \pm\sqrt{2}.$$

The corresponding values of z are 0 and -4. At $O(0, 0, 0)$, $\dfrac{\partial^2 z}{\partial x^2} = 8$, $\dfrac{\partial^2 z}{\partial x \, \partial y} = 0$, $\dfrac{\partial^2 z}{\partial y^2} = -8$. Since

$$\left(\dfrac{\partial^2 z}{\partial x \, \partial y}\right)^2 > \dfrac{\partial^2 z}{\partial x^2} \cdot \dfrac{\partial^2 z}{\partial y^2},$$

it follows that O is a saddle point.

At P, $(0, \sqrt{2}, -4)$, $\dfrac{\partial^2 z}{\partial x^2} = 12$, $\dfrac{\partial^2 z}{\partial x \, \partial y} = 0$, $\dfrac{\partial^2 z}{\partial y^2} = 16$. Since

$$\left(\dfrac{\partial^2 z}{\partial x \, \partial y}\right)^2 < \dfrac{\partial^2 z}{\partial x^2} \cdot \dfrac{\partial^2 z}{\partial y^2} \quad \text{and} \quad \dfrac{\partial^2 z}{\partial x^2} > 0,$$

it follows that P is a minimum point. Similarly $(0, -\sqrt{2}, -4)$ is a minimum point.

The section of the surface by the plane XOZ is shown in Fig. 6.1; the section by the plane YOZ is shown in Fig. 6.2.

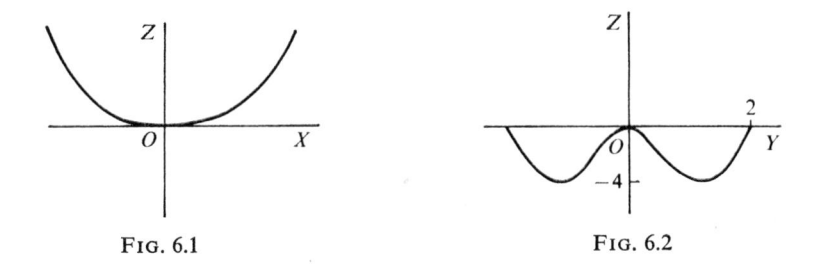

FIG. 6.1

FIG. 6.2

It is clear that the least value of z in the given region is the minimum value -4. The greatest value occurs when $x = 2$, $y = 0$ and is 16. We note that the greatest and least values of z in a restricted x, y domain will occur at either maximum or minimum points within the domain or at points on the boundary of the domain.

Exercises 6(a)

1. Prove that $z = x^2 y^2 (x^2 + y^2 - 1)$ has a maximum value $-\dfrac{1}{27}$ for four sets of values of x and y and a minimum value at $x = 0$, $y = 0$.
2. Investigate the maximum and minimum values of
$$(3u^2 - 2uv + 3v^2) e^{-2u^2 - 2v^2}.$$
3. Prove that the maximum value of $\sin x + \sin y + \sin(x + y)$ is $3\sqrt{3}/2$.

4. Show that the surface $z = x^2y(2y-x+8)$ has a minimum point $(4, -1, -32)$.

5. P is a variable point lying within a given triangle ABC of area Δ. Prove that the minimum value of the sum of the squares of the perpendiculars drawn from P to the sides of the triangle is $4\Delta^2/(a^2+b^2+c^2)$.

6. Investigate the stationary values of $f(x, y) = y^2(y^2-1)+x^2(y^2+1)$. Also find the greatest and least values of $f(x, y)$ in the region $x^2+y^2 \leq 1$.

7. P is a variable point lying within a given triangle ABC. Find the position of P when the sum of the squares of the distances of P from the three vertices of the triangle is a minimum and show that this position is the mean centre of the triangle. Is the value at the mean centre the least value of $PA^2+PB^2+PC^2$?

8. Find the stationary values of the function $y(x^2+y^2-4)$ in the region defined by $x \geq 0$, $x^2+y^2 \leq 9$. Also find the least and greatest values attained by the function in the region.

9. Find the stationary values of the function $(2x+y)(x^2+y^2+2y)$ in the region defined by $x^2+y^2 \leq 4$ and investigate the nature of these values. Find also the greatest and least values attained by the function in this region.

Maximum and minimum values of $V = f(x, y, z)$ when there are constraints on the variables x, y, z

If there is a single constraint on x, y, z, say $\phi(x, y, z) = 0$, then V is essentially a function of two independent variables; if there are two constraints, say $\phi(x, y, z) = 0$, $F(x, y, z) = 0$, then V is essentially a function of a single variable. We shall illustrate two techniques for dealing with such problems.

Example. The squares of the lengths of the principal semi-axes of the ellipse in which the plane

$$2x+y+z = 0 \qquad (1)$$

cuts the ellipsoid

$$x^2+3y^2+3z^2 = 1 \qquad (2)$$

are the maximum and minimum values of

$$V = x^2+y^2+z^2. \qquad (3)$$

At stationary values of V, $\delta V = 0$, i.e.

$$2x\,\delta x+2y\,\delta y+2z\,\delta z = 0.$$

But, from eqn (1),

$$2\delta x+\delta y+\delta z = 0,$$

VALUES OF A FUNCTION OF TWO VARIABLES

and from eqn (2),
$$2x\,\delta x + 6y\,\delta y + 6z\,\delta z = 0.$$

Eliminating $\delta x, \delta y, \delta z$ from the last three equations, we have

$$\begin{vmatrix} 2x & 2y & 2z \\ 2 & 1 & 1 \\ 2x & 6y & 6z \end{vmatrix} = 0,$$

which gives $x(y-z) = 0$. Hence, the stationary values of V occur when $x = 0, y = \pm 1/\sqrt{6}, z = \mp 1/\sqrt{6}$; and $x = \pm 1/\sqrt{7}, y = \mp 1/\sqrt{7}, z = \mp 1/\sqrt{7}$. It is clear geometrically that these points give maximum and minimum values of V and it follows that the principal semi-axes of the ellipse are $1/\sqrt{3}$ and $\sqrt{\frac{3}{7}}$.

Example. Find the maximum and minimum values of

$$V = xyz, \tag{1}$$

subject to the condition

$$x^2 + y^2 + z^2 = 1. \tag{2}$$

V is essentially a function of two independent variables which we shall take as x and y. At stationary values of V, $\delta V = 0$, i.e.

$$yz\,\delta x + zx\,\delta y + xy\,\delta z = 0, \tag{3}$$

while, from eqn. (2),

$$2x\,\delta x + 2y\,\delta y + 2z\,\delta z = 0. \tag{4}$$

On eliminating δz from eqns (3) and (4), we obtain

$$(yz^2 - yx^2)\,\delta x + (xz^2 - xy^2)\,\delta y = 0.$$

Now, since x and y are independent, the coefficients of δx and δy in the last equation must be zero, i.e.

$$yz^2 - yx^2 = 0 \tag{5}$$

and

$$xz^2 - xy^2 = 0. \tag{6}$$

From eqns (5), (6) and (2) we obtain the following solutions: $x = 0$, $y = 0$, $z = \pm 1$; $x = 0$, $y = \pm 1$, $z = 0$; $x = \pm 1$, $y = 0$, $z = 0$; $\pm x = \pm y = \pm z = 1/\sqrt{3}$.

By examining the values of V in the neighbourhood of these points, it is easy to see that there are maximum values at the points

$\pm x = \pm y = \pm z = 1/\sqrt{3}$, when all or one of the ambiguous signs is taken as positive. The maximum value is $1/(3\sqrt{3})$ in each case.

There are minimum values at the points $\pm x = \pm y = \pm z = 1/\sqrt{3}$, when all or one of the ambiguous signs is taken as negative. The minimum value is $-1/(3\sqrt{3})$ in each case.

Maximum and minimum values by Lagrange's method of undetermined multipliers

An interesting method of dealing with problems of conditional maxima and minima was devised by Lagrange and is called the method of undetermined multipliers. It is illustrated in the following problem.

Example. Find the values of x, y, z for which the function

$$V = xyz \tag{1}$$

is stationary subject to the condition

$$\frac{x^2}{a^2} + \frac{y^2}{b^2} + \frac{z^2}{c^2} = 1. \tag{2}$$

$$\delta V = yz\,\delta x + xz\,\delta y + xy\,\delta z, \tag{3}$$

while, from eqn (2)

$$\frac{x}{a^2}\delta x + \frac{y}{b^2}\delta y + \frac{z}{c^2}\delta z = 0. \tag{4}$$

It follows that, for all values of λ,

$$\delta V = \left(yz - \lambda\frac{x}{a^2}\right)\delta x + \left(zx - \lambda\frac{y}{b^2}\right)\delta y + \left(xy - \lambda\frac{z}{c^2}\right)\delta z.$$

Now choose λ so that the coefficient of δz is zero. Then, since δx and δy are independent, their coefficients must also be zero for stationary values of V. We are thus led to the equations

$$yz = \lambda\frac{x}{a^2}; \qquad zx = \lambda\frac{y}{b^2}: \qquad xy = \lambda\frac{z}{c^2}.$$

From these equations

$$(yz)(zx)(xy) = \lambda^3\frac{xyz}{a^2b^2c^2},$$

whence
$$xyz = \frac{\lambda^3}{a^2b^2c^2}.$$
Also
$$xyz = \frac{\lambda x^2}{a^2} = \frac{\lambda y^2}{b^2} = \frac{\lambda z^2}{c^2},$$
and hence from eqn (2), $3xyz = \lambda$. It follows that
$$\frac{\lambda}{3} = \frac{\lambda^3}{a^2b^2c^2} \quad \text{and} \quad \lambda = \frac{abc}{\sqrt{3}}, \, V = \frac{abc}{3\sqrt{3}}.$$
It is clear that this value of V is a maximum value.

Example. Find the least and greatest distances from the point $(0, 1, 1)$ to the circle of intersection of the sphere
$$x^2+y^2+z^2 = 1 \tag{1}$$
and the plane
$$2x+y+z = 2. \tag{2}$$
$$V = x^2+(y-1)^2+(z-1)^2,$$
where x, y, z satisfy (1) and (2). V is essentially a function of a single variable, and can be so expressed, but it is simpler to proceed as follows.
$$\delta V = 2x \, \delta x + 2(y-1) \, \delta y + 2(z-1) \, \delta z,$$
where, from (1),
$$2x \, \delta x + 2y \, \delta y + 2z \, \delta z = 0,$$
and, from (2),
$$2\delta x + \delta y + \delta z = 0.$$
We use two Lagrange multipliers λ and μ, and deduce that
$$\delta V = (2x+2x\lambda+2\mu) \, \delta x + (2y-2+2y\lambda+\mu) \, \delta y + (2z-2+2z\lambda+\mu) \, \delta z$$
Hence, for maximum and minimum points,
$$2x+2x\lambda+2\mu = 0,$$
$$2y-2+2y\lambda+1\mu = 0,$$
$$2z-2+2z\lambda+1\mu = 0.$$
Eliminating λ and μ, we have
$$\begin{vmatrix} 2x & 2x & 2 \\ 2y-2 & 2y & 1 \\ 2z-2 & 2z & 1 \end{vmatrix} = 0. \tag{3}$$

Solving eqns (1), (2) and (3), we obtain $x = 1$, $y = 0$, $z = 0$; $x = \frac{1}{3}$, $y = \frac{2}{3}$, $z = \frac{2}{3}$. Clearly, the maximum value of V is 3 and the minimum value $\frac{1}{3}$. It follows that the greatest distance is $\sqrt{3}$, and the least distance $1/\sqrt{3}$.

Exercises 6(b)

1. Find the lengths of the principal axes of the section of the ellipsoid $x^2 + 2y^2 + 3z^2 = 105$ by the plane $2x - y + 2z = 0$.

2. Find the maximum value of $x^p y^q$, where $ax + by = c$, given that p, q, a, b, c are positive constants.

3. The variables x, y, z are connected by the relation $xy + yz + zx = 3k^2$, where k is a positive constant. Find the values of x, y, z for which xyz is stationary.

4. If $\phi(x) + \phi(y) + \phi(z) = 3c$, prove that the function $f(x) + f(y) + f(z)$ will be stationary when $x = y = z = k$, where $\phi(k) = c$. Show further that the function will be a maximum or minimum according as

$$\frac{f''(k)\phi'(k) - f'(k)\phi''(k)}{\phi'(k)}$$

is positive or negative.

5. If $V = px + qy + rz$, where

$$ax^2 + by^2 + cz^2 + 2fyz + 2gzx + 2hxy = 1,$$

prove that the maximum and minimum values of v satisfy the equation

$$\begin{vmatrix} a & h & g & p \\ h & b & f & q \\ g & f & c & r \\ p & q & r & v^2 \end{vmatrix} = 0.$$

Directional derivatives

A space curve is defined as the set of points $P(x, y, z)$ for which x, y, z are given functions of a parameter t, e.g. the equations of a circular helix are $x = a\cos t$, $y = a\sin t$, $z = bt$. If s denotes the length of the arc AP on the curve, measured from a fixed point A on the curve, then s is a suitable parameter and we may consider the equation of the space curve in the form

$$x = f_1(s), \qquad y = f_2(s), \qquad z = f_3(s).$$

Let $P(x, y, z)$ and $Q(x + \delta x, y + \delta y, z + \delta z)$ be two neighbouring points on the curve, and denote the arc PQ by δs.

Then if $f(x, y, z)$ denotes any function of x, y, z and

$$\lim_{\delta s \to 0} \frac{f(x+\delta x, y+\delta y, z+\delta z) - f(x, y, z)}{\delta s}$$

exists, this limit is called the directional derivative of $f(x, y, z)$ along the curve and is denoted by df/ds.

It follows that the directional derivative is

$$\frac{df}{ds} = \frac{\partial f}{\partial x}\frac{dx}{ds} + \frac{\partial f}{\partial y}\frac{dy}{ds} + \frac{\partial f}{\partial z}\frac{dz}{ds}.$$

If the equations of the space curve are given in terms of a parameter t, other than s, then we may write the formula for the directional derivative in the form

$$\frac{df}{ds} = \frac{\partial f}{\partial x}\frac{dx}{dt}\frac{dt}{ds} + \frac{\partial f}{\partial y}\frac{dy}{dt}\frac{dt}{ds} + \frac{\partial f}{\partial z}\frac{dz}{dt}\frac{dt}{ds},$$

where

$$\left(\frac{ds}{dt}\right)^2 = \left(\frac{dx}{dt}\right)^2 + \left(\frac{dy}{dt}\right)^2 + \left(\frac{dz}{dt}\right)^2.$$

If **i**, **j**, **k** denote unit vectors along the coordinate axes, and **r** denotes the position vector **OP**, then

$$\mathbf{r} = x\mathbf{i} + y\mathbf{j} + z\mathbf{k}.$$

For any given function f, the vector $\frac{\partial f}{\partial x}\mathbf{i} + \frac{\partial f}{\partial y}\mathbf{j} + \frac{\partial f}{\partial z}\mathbf{k}$ is called gradient f, or grad f.

It is clear that we may write the directional derivative in the form of a scalar product

$$\left(\frac{\partial f}{\partial x}\mathbf{i} + \frac{\partial f}{\partial y}\mathbf{j} + \frac{\partial f}{\partial z}\mathbf{k}\right) \cdot \left(\frac{dx}{ds}\mathbf{i} + \frac{dy}{ds}\mathbf{j} + \frac{dz}{ds}\mathbf{k}\right) = (\text{grad} f) \cdot \left(\frac{d\mathbf{r}}{ds}\right).$$

Now, the direction cosines of the tangent to the curve at P are $\frac{dx}{ds}, \frac{dy}{ds}, \frac{dz}{ds}$ and hence the unit vector along the tangent at P is $\frac{dx}{ds}\mathbf{i} + \frac{dy}{ds}\mathbf{j} + \frac{dz}{ds}\mathbf{k}$. Denote this unit vector by **t**. Then, the directional derivative $\frac{df}{ds} = (\text{grad} f) \cdot (\mathbf{t})$.

Differentiations of integrals containing a parameter

If $f(x, t)$ is an integrable function of x and $\dfrac{\partial f}{\partial t}$ a continuous function of x and t, then

$$\frac{d}{dt}\int_a^b f(x, t)\,dx = \int_a^b \frac{\partial f}{\partial t}\,dx,$$

where a and b are constants. For, if we denote $\int_a^b f(x, t)\,dx$ by $\phi(t)$,

$$\frac{\phi(t+\delta t) - \phi(t)}{\delta t} = \int_a^b \frac{f(x, t+\delta t) - f(x, t)}{\delta t}\,dx$$

$$= \int_a^b \frac{\partial}{\partial t} f(x, t+\theta\,\delta t)\,dx,$$

where $0 < \theta < 1$, by the first mean value theorem. Now

$$\int_a^b \frac{\partial}{\partial t} f(x, t+\theta\,\delta t)\,dx$$

tends to $\int_a^b \dfrac{\partial}{\partial t} f(x, t)\,dx$ as $\delta t \to 0$, since $\dfrac{\partial f}{\partial t}$ is a continuous function of t. Hence $\phi'(t) = \int_a^b \dfrac{\partial f}{\partial t}\,dx$, which is the required result.

Euler's theorem on homogeneous functions

The name of the distinguished Swiss mathematician Leonard Euler (1707–1783) is associated with the following theorem.

If z is a homogeneous polynomial of degree n in x and y,

$$ax^n + bx^{n-1}y + cx^{n-2}y^2 + \cdots = x^n\left\{a + b\left(\frac{y}{x}\right) + c\left(\frac{y}{x}\right)^2 + \cdots\right\},$$

then z may be expressed in the form $z = x^n f\left(\dfrac{y}{x}\right)$, where $f\left(\dfrac{y}{x}\right)$ is a polynomial in $\dfrac{y}{x}$. More generally, we shall refer to any function of the form $z = x^n f\left(\dfrac{y}{x}\right)$, where f is not necessary a polynomial, e.g.

$$f\left(\frac{y}{x}\right) = \sin\left(\frac{y}{x}\right), \quad \frac{y}{x}e^{y/x}, \quad \frac{x^3}{y^3}\log\left(2+\frac{y}{x}\right),$$

VALUES OF A FUNCTION OF TWO VARIABLES

as a homogeneous function of degree n in x and y. It will be clear that if $\phi(x, y)$ is a homogeneous function of degree n in x and y, then

$$\phi(kx, ky) = k^n \phi(x, y).$$

Euler's theorem states that if z is a homogeneous function of degree n in x and y, then

$$x \frac{\partial z}{\partial x} + y \frac{\partial z}{\partial y} = nz.$$

To prove this theorem we write z in the form $z = x^n f\left(\frac{y}{x}\right)$, and differentiate partially with respect to x and y, giving

$$\frac{\partial z}{\partial x} = x^n f'\left(\frac{y}{x}\right)\left(-\frac{y}{x^2}\right) + nx^{n-1} f\left(\frac{y}{x}\right),$$

and

$$\frac{\partial z}{\partial y} = x^n f'\left(\frac{y}{x}\right)\left(\frac{1}{x}\right).$$

Hence,

$$x \frac{\partial z}{\partial x} + y \frac{\partial z}{\partial y} = nx^n f\left(\frac{y}{x}\right) = nz.$$

The converse of the theorem is also true, namely, that if

$$x \frac{\partial z}{\partial x} + y \frac{\partial z}{\partial y} = nz,$$

then z is a homogenous function of x and y. For we can express *any* function of x and y in the form $z = x^n \phi(x, u)$, where $u = y/x$. Differentiating partially with respect to x, y, we have

$$\frac{\partial z}{\partial x} = x^n \left(\frac{\partial \phi}{\partial x} + \frac{\partial \phi}{\partial u} \frac{\partial u}{\partial x}\right) + nx^{n-1} \phi(x, u)$$

$$= x^n \left\{\frac{\partial \phi}{\partial x} + \frac{\partial \phi}{\partial u}\left(-\frac{y}{x^2}\right)\right\} + nx^{n-1} \phi(x, u),$$

$$\frac{\partial z}{\partial y} = x^n \frac{\partial \phi}{\partial u} \frac{\partial u}{\partial y} = x^n \frac{\partial \phi}{\partial u}\left(\frac{1}{x}\right).$$

On substituting these values of $\frac{\partial z}{\partial x}$ and $\frac{\partial z}{\partial y}$ in the given equation

$$x \frac{\partial z}{\partial x} + y \frac{\partial z}{\partial y} = nz,$$

we get $\dfrac{\partial \phi}{\partial x} = 0$. It follows that the function $\phi(x, u)$ is a function of u only, and hence that z is a homogeneous function of x and y.

Exercises 6(c)

1. Prove that if z is a homogeneous function of degree n in x and y,
$$x^2\dfrac{\partial^2 z}{\partial x^2} + 2xy\dfrac{\partial^2 z}{\partial x \partial y} + y^2\dfrac{\partial^2 z}{\partial y^2} = n(n-1)z,$$
and that
$$x^3\dfrac{\partial^3 z}{\partial x^3} + 3x^2 y\dfrac{\partial^3 z}{\partial x^2 \partial y} + 3xy^2\dfrac{\partial^3 z}{\partial x \partial y^2} + y^3\dfrac{\partial^3 z}{\partial y^3} = n(n-1)(n-2)z.$$

2. Find the directional derivative of xy^2z^3 at the point $(1, 2, 1)$ on the curve $x = t$, $y = t^2 + 1$, $z = t^3$.

3. (i) Prove that the maximum directional derivative of $f(x, y, z)$ is in the direction of grad f.
 (ii) Prove that grad f is normal to the curve $f(x, y, z) = c$.

4. If $\phi(t) = \displaystyle\int_0^1 \dfrac{1}{\log x}(x^t - 1)\, dx$, find $\phi'(t)$ and hence evaluate the integral.

5. If $F(x, y) = f(x, y) + g(x, y)$, where f is a homogeneous polynomial of degree n in x and y, and g is a homogeneous polynomial of degree m in x and y, prove that
$$x^2\dfrac{\partial^2 F}{\partial x^2} + 2xy\dfrac{\partial^2 F}{\partial x \partial y} + y^2\dfrac{\partial^2 F}{\partial y^2} - (m-1)\left(x\dfrac{\partial F}{\partial x} + y\dfrac{\partial F}{\partial y}\right) = n(n-m)f(x, y).$$

Miscellaneous problems on Chapter 6

1. $f(x, y)$ is a homogeneous function of x and y of degree n. Show that the points of inflexion of the curve $f(x, y) = 1$ lie on the curve
$$(n-1)\dfrac{\partial f}{\partial x}\dfrac{\partial f}{\partial y} = n\dfrac{\partial^2 f}{\partial x \partial y}.$$

2. Show that $\dfrac{x^n + y^n + 1}{(x+y+1)^n}$ has a maximum value at $x = 1$, $y = 1$ if $0 < n < 1$. Discuss what happens if n lies outside this range.

3. Investigate the maximum and minimum values of z where
$$z = xy^2(2x - y + 8).$$

4. If $\mathbf{i}, \mathbf{j}, \mathbf{k}$ denote unit vectors along the coordinate axes, find a unit vector along the normal at any point of the following surfaces:
  (i) $f(x, y, z) = 0$, (ii) $z = \phi(x, y)$, (iii) $\left.\begin{array}{l}f(x, y, z, t) = 0 \\ \phi(x, y, z, t) = 0\end{array}\right\}.$

5. x_{rs} denotes the element in the rth row and sth column of the determinant Δ of order n. The n^2 variables x_{rs} are subject to the conditions

$$\sum_{s=1}^{n} (x_{rs})^2 = (k_r)^2, \quad k_r > 0,$$

for all values of r, $1 \leq r \leq n$. Use Lagrange's method of undetermined multipliers to show that for stationary values of Δ,

$$X_{rs} = \lambda_r x_{rs},$$

where X_{rs} is the cofactor of x_{rs} and λ_r depends on r only, not on s.
Deduce that

$$\sum_{i=1}^{n} x_{ri} x_{ti} = 0 \quad \text{if} \quad r \neq t.$$

Hence, prove that Δ^2 is a determinant with terms in the leading diagonal $k_1^2, k_2^2, \ldots$ and all other terms zero, and hence that a stationary value of Δ is

$$k_1 k_2 k_3 \cdots k_n.$$

Prove that this stationary value is a maximum.

6. P is a point on the plane $x+y+z = 3p$, Q a point on the ellipsoid $x^2/a^2 + y^2/b^2 + z^2/c^2 = 1$. Find the minimum value of PQ.

7. R is the semicircular region defined by $x^2+y^2 \leq 1$, $y \geq 0$. Find the maximum and minimum values, if any, of the following functions within R, and consider the behaviour of the function on the boundary of R. Hence find the greatest and least value of the function attained in R.
(i) $x+2y$, (ii) $(x+2y)^2$, (iii) xy, (iv) $x(x^2+y^2-1)$,
(v) $2x^2+3y^2+2x-2y$, (vi) $xy^2(x^2+2y^2-1)$.

8. If x, y, z are positive and satisfy $x+y+z = 1$, prove that the maximum value of $x^a y^b z^c$, where a, b, c are positive constants, is $a^a b^b c^c/(a+b+c)^{a+b+c}$.

9. x, y, z are independent variables taking only positive values; p, q, r are positive constants satisfying $p \leq \frac{1}{2}$, $q \leq \frac{1}{2}$, $r \leq \frac{1}{2}$,

$$p+q+r = 1.$$

Prove that $px+qy+rz-x^p y^q z^r$ has stationary values when $x = t$, $y = t$, $z = t$ for all values of t. By investigating the value of the function when $x = t+\alpha$, $y = t+\beta$, $z = t+\gamma$, where α, β, γ are small, investigate the nature of the stationary values. Hence prove that

$$px+qy+rz \geq x^p y^q z^r.$$

Show that the arithmetic mean of x, y, z is greater than their geometric mean. Generalize this result and prove the general theorem.

10. $V = f(x, y, z)$, where the variables x, y, z are subject to the restriction $g(x, y, z) = 0$, f and g being differentiable functions of x, y, z.

If P is a point for which V has a stationary value V_0 prove that the surfaces $f(x, y, z) = V_0$ and $g(x, y, z) = 0$ touch at P, assuming that P is not a singular point on either surface.

11. $z = f(x, y)$ is transformed into $z = \phi(u, v)$ by the substitution $x = 2uv$, $y = u^2 - v^2$. Prove that

$$4(u^2 + v^2)\left(\frac{\partial^2 f}{\partial x^2} + \frac{\partial^2 f}{\partial y^2}\right) = \frac{\partial^2 \phi}{\partial u^2} + \frac{\partial^2 \phi}{\partial v^2}.$$

12. Find the maximum value of $V = x^a y^b z^c$, where x, y, z are positive variables, a, b, c are positive constants, the variables x, y, z being subject to the relation $x + y + z = k$, where k is a constant.

Hence, or otherwise, prove that for all positive values of x, y, z

$$x^{\frac{1}{8}} y^{\frac{3}{8}} z^{\frac{1}{2}} \leq \tfrac{1}{8}x + \tfrac{3}{8}y + \tfrac{1}{2}z.$$

13. Find the values of x, y, z, t, for which $V = x^3 y^2 z t$ has stationary values, where x, y, z, t are positive variables subject to the constraint

$$x + y + z + t = 1.$$

Hence show that $V \leq \dfrac{108}{7^7}$.

7

Double Integrals

We begin by considering the problem of finding by a calculus method the area of the circle

$$(x-2)^2 + (y-3)^2 = 1.$$

For any value of x between 1 and 3, there are two values of y, y_1 and y_2 (see Fig. 7.1), where

$$y_1 = 3 - \sqrt{\{1-(x-2)^2\}},$$
$$y_2 = 3 + \sqrt{\{1-(x-2)^2\}}.$$

It is clear that the area of the circle is

$$A = \int_1^3 y_2\, dx - \int_1^3 y_1\, dx$$

$$= \int_1^3 [3+\sqrt{\{1-(x-2)^2\}}]\, dx - \int_1^3 [3-\sqrt{\{1-(x-2)^2\}}]\, dx$$

$$= 2\int_1^3 \sqrt{\{1-(x-2)^2\}}\, dx,$$

which, on writing $x-2 = \sin\theta$, $\theta = \text{P.V. } \sin^{-1}(x-2)$, gives

$$A = 2\int_{-\pi/2}^{\pi/2} (\cos\theta)(\cos\theta)\, d\theta = 4\int_0^{\pi/2} \cos^2\theta\, d\theta = \pi.$$

The procedure may be simplified by using a parametric equation of the circle instead of the Cartesian equation (see Fig. 7.2). We select as parameter t the angle made by the radius CP and CE, parallel to the x-axis.

Then
$$x = 2 + \cos t,$$
$$y = 3 + \sin t.$$

174   MATHEMATICAL ANALYSIS AND TECHNIQUES

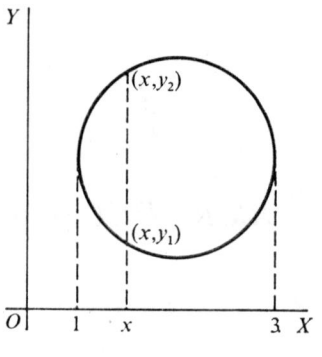

FIG. 7.1

As the point $P(x, y)$ describes the circle in an anticlockwise direction from E to E, t steadily increases from 0 to 2π.

$$A = \int_1^3 y_2 \, dx - \int_1^3 y_1 \, dx$$

$$= \int_\pi^0 y_2 \frac{dx}{dt} dt - \int_\pi^{2\pi} y_1 \frac{dx}{dt} dt$$

$$= \int_\pi^0 (3+\sin t)\frac{dx}{dt} dt - \int_\pi^{2\pi} (3+\sin t)\frac{dx}{dt} dt.$$

Note that y_1 and y_2 are different functions of x but the same

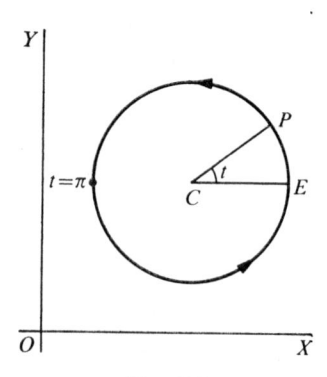

FIG. 7.2

function of t. Hence

$$A = \int_\pi^0 - \int_\pi^{2\pi} (3+\sin t)(-\sin t)\,dt$$

$$= \int_0^{2\pi} (3+\sin t)\sin t\,dt = \pi.$$

We shall now extend the procedure of using parametric equations to find the area of any simple closed curve.

Area of a closed curve

Consider a closed curve lying in the first quadrant (see Fig. 7.3) such that, for any value of x between a and b, the ordinate at x cuts the curve in two points (x, y_1) and (x, y_2), where $y_2 > y_1$.

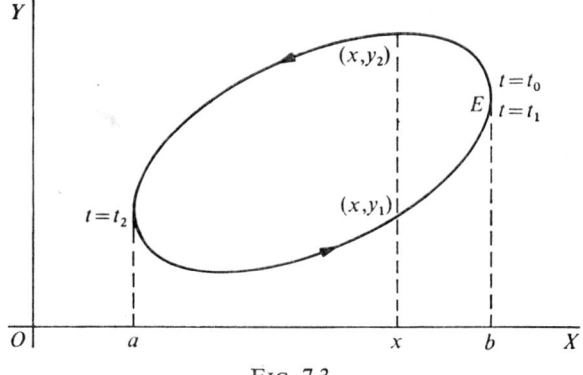

Fig. 7.3

Then the area of the curve is given by

$$A = \int_a^b y_2\,dx - \int_a^b y_1\,dx.$$

If we can find a parameter t such that x and y are single-valued differentiable functions of t, $x = f(t)$, $y = \phi(t)$, and as the point $P(x, y)$ moves round the curve in an anticlockwise direction from E to E, t steadily increases or steadily decreases from t_0 to t_1, then

$$A = \int_{t_2}^{t_0} y_2 \frac{dx}{dt}\,dt - \int_{t_2}^{t_1} y_1 \frac{dx}{dt}\,dt,$$

where t_2 is the value of t at $x = a$. Now y_1 and y_2 are different functions of x but the same function of t, and in these integrals in t, we can replace y_1 and y_2 by $\phi(t)$ or y. Hence

$$A = \int_{t_2}^{t_0} y\frac{dx}{dt}dt - \int_{t_2}^{t_1} y\frac{dx}{dt}dt$$

$$= -\int_{t_0}^{t_2} - \int_{t_2}^{t_1} y\frac{dx}{dt}dt$$

$$= -\int_{t_0}^{t_1} y\frac{dx}{dt}dt.$$

Alternatively (see Fig. 7.4), if we denote the values of t at $y = c$, $y = d$ by t_3, t_4 respectively,

$$A = \int_c^d x_2\,dy - \int_c^d x_1\,dy$$

$$= \left\{\int_{t_3}^{t_1} + \int_{t_0}^{t_4} x_2\frac{dy}{dt}dt\right\} - \int_{t_3}^{t_4} x_1\frac{dy}{dt}dt$$

$$= \int_{t_3}^{t_1} + \int_{t_0}^{t_4} + \int_{t_4}^{t_3} x\frac{dy}{dt}dt$$

$$= \int_{t_0}^{t_1} x\frac{dy}{dt}dt.$$

The restrictions that the curve lies in the first quadrant and that any ordinate cuts the curve in only two points are not necessary, and the results are valid for any simple closed curve in the xy-plane, for example, for the curve shown in Fig. 7.5, but not for a curve with a node as in Fig. 7.6. In the latter case the integral would yield the difference in area of the two loops.

We shall apply the results

$$A = -\int_{t_0}^{t_1} y\frac{dx}{dt}dt, \qquad A = \int_{t_0}^{t_1} x\frac{dy}{dt}dt,$$

DOUBLE INTEGRALS

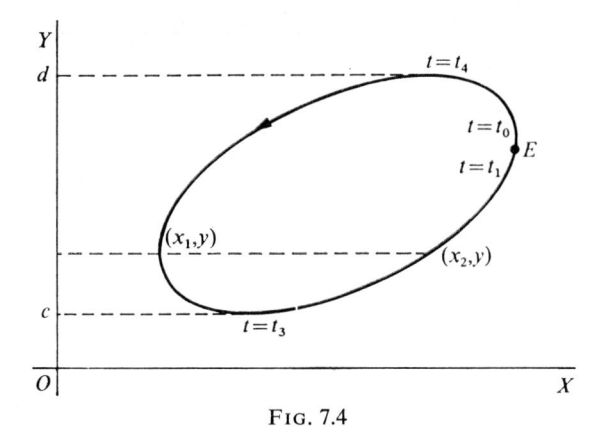

FIG. 7.4

to find two independent solutions of the area of the circle

$$(x-3)^2 + (y-4)^2 = 4.$$

For we can select a parameter t, such that

$$x = 3 + 2\cos t,$$
$$y = 4 + 2\sin t;$$

as the point $P(x, y)$ moves round the circle in an anticlockwise direction starting from the point $(5, 4)$, t steadily increases from 0 to 2π. Thus

$$A = -\int_0^{2\pi} y \frac{dx}{dt} dt = -\int_0^{2\pi} (4 + 2\sin t)(-2\sin t) \, dt = 4\pi$$

or

$$A = \int_0^{2\pi} x \frac{dy}{dt} dt = \int_0^{2\pi} (3 + 2\cos t)(2\cos t) \, dt = 4\pi.$$

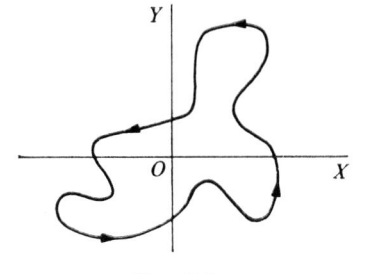

FIG. 7.5

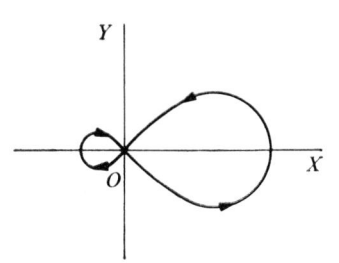

FIG. 7.6

Either of the results for the area of a simple closed curve

$$A = -\int_{t_0}^{t_1} y\frac{dx}{dt}dt, \quad A = \int_{t_0}^{t_1} x\frac{dy}{dt}dt,$$

can be obtained from the other using the method of integration by parts. For

$$A = -\int_{t_0}^{t_1} y\frac{dx}{dt}dt = [-yx]_{t_0}^{t_1} + \int_{t_0}^{t_1} x\frac{dy}{dt}dt = \int_{t_0}^{t_1} x\frac{dy}{dt}dt.$$

From the two standard results for A, we deduce a third result which sometimes leads to a simpler integral for A, viz.

$$A = \frac{1}{2}\int_{t_0}^{t_1}\left(x\frac{dy}{dt} - y\frac{dx}{dt}\right)dt.$$

If, as t steadily increases or steadily decreases from t_0 to t_1, the point $P(x, y)$ moves round the curve in a *clockwise* direction, then the three results stated above must be modified. It is left to the reader to prove that, in this case,

$$A = \int_{t_0}^{t_1} y\frac{dx}{dt}dt; \quad A = -\int_{t_0}^{t_1} x\frac{dy}{dt}dt; \quad A = \frac{1}{2}\int_{t_0}^{t_1}\left(y\frac{dx}{dt} - x\frac{dy}{dt}\right)dt.$$

It is usual to say that a closed curve is described positively when it is described in an anticlockwise direction. In the proofs given above, the values t_0 and t_1 of the parameter t refer to the point E, but the results are clearly true for any starting point A as in Fig. 7.7. The proof of this fact is left as an exercise for the reader.

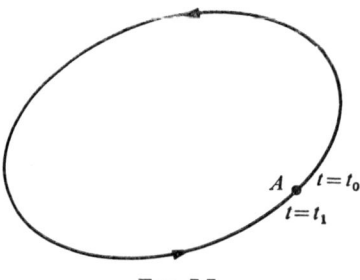

FIG. 7.7

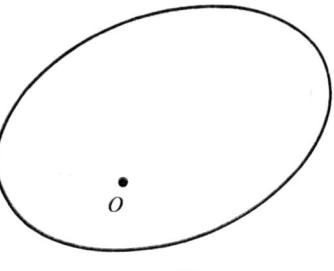

FIG. 7.8

Area of a simple closed curve whose equation is given in polar coordinates

Two cases arise. If the origin lies within the curve (Fig. 7.8) r is a single-valued function of θ and the area of the curve is

$$A = \int_0^{2\pi} \tfrac{1}{2} r^2 \, d\theta.$$

If, on the other hand, the origin lies outside the curve (Fig. 7.9) then for every value of θ between α and β there are two values of r, r_1 and r_2, and

$$A = \int_\alpha^\beta \tfrac{1}{2} r_2^2 \, d\theta - \int_\alpha^\beta \tfrac{1}{2} r_1^2 \, d\theta.$$

We can use the same parameter t as in the Cartesian form of the

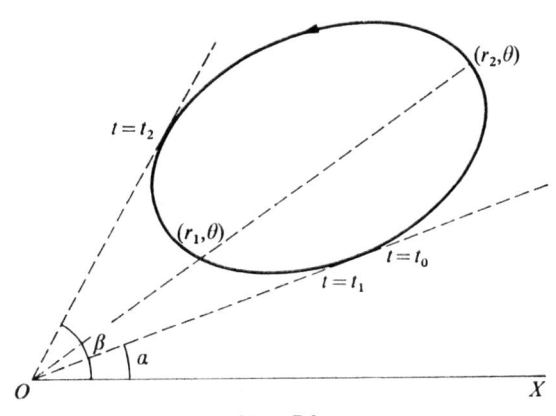

FIG. 7.9

curve. Then, in the first case

$$A = \int_{t_0}^{t_1} \tfrac{1}{2}r^2 \frac{d\theta}{dt} dt.$$

In the second case,

$$A = \int_{t_0}^{t_2} \tfrac{1}{2}r_2^2 \frac{d\theta}{dt} dt - \int_{t_1}^{t_2} \tfrac{1}{2}r_1^2 \frac{d\theta}{dt} dt$$

$$= \int_{t_0}^{t_2} \tfrac{1}{2}r^2 \frac{d\theta}{dt} dt - \int_{t_1}^{t_2} \tfrac{1}{2}r^2 \frac{d\theta}{dt} dt$$

$$= \int_{t_0}^{t_2} + \int_{t_2}^{t_1} \tfrac{1}{2}r^2 \frac{d\theta}{dt} dt$$

$$= \int_{t_0}^{t_1} \tfrac{1}{2}r^2 \frac{d\theta}{dt} dt.$$

It is easy to prove directly that the results

$$A = \frac{1}{2} \int_{t_0}^{t_1} \left(x \frac{dy}{dt} - y \frac{dx}{dt} \right) dt,$$

$$A = \frac{1}{2} \int_{t_0}^{t_1} r^2 \frac{d\theta}{dt} dt,$$

for the area of a simple closed curve are equivalent. For, from the relations $x = r \cos \theta$, $y = r \sin \theta$, we have

$$\frac{dx}{dt} = -r \sin \theta \frac{d\theta}{dt} + \cos \theta \frac{dr}{dt},$$

$$\frac{dy}{dt} = r \cos \theta \frac{d\theta}{dt} + \sin \theta \frac{dr}{dt}.$$

Hence $x \dfrac{dy}{dt} - y \dfrac{dx}{dt} = r^2 \dfrac{d\theta}{dt}$, which proves the result. Alternatively, from $\theta = \tan^{-1} \dfrac{y}{x}$,

$$\frac{d\theta}{dt} = \frac{1}{1+\left(\dfrac{y}{x}\right)^2} \cdot \frac{x\dfrac{dy}{dt} - y\dfrac{dx}{dt}}{x^2}$$

$$= \frac{1}{r^2}\left(x\frac{dy}{dt} - y\frac{dx}{dt}\right)$$

or

$$r^2\frac{d\theta}{dt} = x\frac{dy}{dt} - y\frac{dx}{dt}.$$

Example. Consider the curve $x^3 + y^3 = 3xy$, whose graph is shown in Fig. 7.10. The line $y = tx$ cuts the curve where

$$x = \frac{3t}{1+t^3}, \qquad y = \frac{3t^2}{1+t^3}.$$

As the point (x, y) moves round the loop of the curve in an anticlockwise direction, t steadily increases from 0 to ∞. It follows that the area of the loop is given by

$$A = \frac{1}{2}\int_0^\infty \left(x\frac{dy}{dt} - y\frac{dx}{dt}\right) dt$$

$$= \frac{1}{2}\int_0^\infty x^2\frac{d}{dt}\left(\frac{y}{x}\right) dt$$

$$= \frac{1}{2}\int_0^\infty \frac{9t^2}{(1+t^3)^2}(1)\, dt$$

$$= \frac{3}{2}\left[\frac{-1}{1+t^3}\right]_0^\infty$$

$$= \frac{3}{2}.$$

Obtain the answer also by using the results

$$A = \int_0^\infty x\frac{dy}{dt}dt, \qquad A = -\int_0^\infty y\frac{dx}{dt}dt.$$

FIG. 7.10

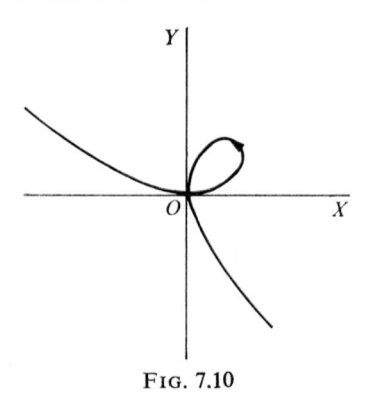

Alternatively, we can write the equation of the curve in polar form

$$r = \frac{3\cos\theta \sin\theta}{(\cos^3\theta + \sin^3\theta)},$$

whence

$$A = \int_0^{\pi/2} \tfrac{1}{2} r^2 \, d\theta = \frac{9}{2}\int_0^{\pi/2} \frac{\cos^2\theta \sin^2\theta}{(\cos^3\theta + \sin^3\theta)^2} d\theta.$$

On writing $\tan\theta = v$,

$$A = \frac{9}{2}\int_0^{\infty} \frac{v^2 \, dv}{(1+v^3)^2} = \frac{3}{2}.$$

Exercises 7(a)

1. Sketch the curve with parametric equations

$$x = t^2 - 1,$$
$$y = t(t^2 - 1),$$

and find the area of the loop of the curve.

2. Sketch the curve whose parametric equations are

$$x = 3a\cos t - a\cos 3t,$$
$$y = 3a\sin t - a\sin 3t,$$

and find the area of the curve.

3. It is clear that the integral $\int_{t_0}^{t_1} x \frac{dy}{dt} dt$ for the area of a closed curve must be independent of the choice of origin. Show this directly by investigating

DOUBLE INTEGRALS

the value of the integral

$$\int_{t_0}^{t_1} (x+h)\frac{d}{dt}(y+k)\,dt,$$

where h and k are constants.

4. Sketch the curve whose parametric equations are

$$x = at(4-t^2), \qquad y = a(4-t^2).$$

Show that the loop is described in a clockwise direction as t steadily increases from -2 to 2, and find the area of the loop.

5. x and y are single-valued functions of t, and as t steadily increases from t_0 to t_1 (x, y) moves along the arc PQ of a curve from P to Q. State in terms of areas what the integral

$$\frac{1}{2}\int_{t_0}^{t_1} \left(x\frac{dy}{dt} - y\frac{dx}{dt}\right) dt$$

represents in Fig. 7.11, Fig. 7.12, Fig. 7.13, and Fig. 7.14.

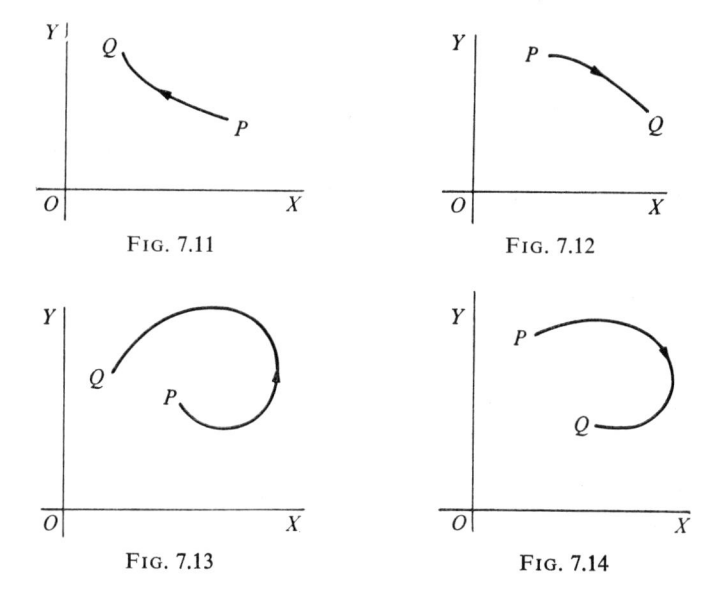

FIG. 7.11

FIG. 7.12

FIG. 7.13

FIG. 7.14

6. A plane curve is given by $x = f(t)$, $y = \phi(t)$, and the point (x, y) moves along the curve from P to Q as t steadily increases from t_0 to t_1. Prove that the volume generated when the sectorial area OPQ is rotated

through 2π about the x-axis is

$$\frac{2\pi}{3} \int_{t_0}^{t_1} y\left(x\frac{dy}{dt} - y\frac{dx}{dt}\right) dt.$$

Prove that the volume generated when the area enclosed by the curve

$$x = a(1-t^2), \quad y = 2a(1+t)$$

and the y-axis is rotated through 2π about the x-axis is five times the volume generated when the same area is rotated about the y-axis.

7. Sketch the curve $x = a\cos t$, $y = a\sin 2t$ and show that it forms two loops of equal size. Find the area enclosed by one of the loops.

Find the equation of the normal at the point on the curve with parameter t and hence or otherwise show that the centre of curvature at the point $(a, 0)$ is the point $(-3a, 0)$.

8. Sketch the curve whose parametric equations are

$$x = (t-1)e^{-t}, \quad y = t(t-1)e^{-t},$$

and find the area of its loop.

Consider the curve $f(x, y) = 0$ whose graph is shown in Fig. 7.15. We wish to find the area bounded by the portion of the curve $PQRST$, and the lines PA, TB, and AB.

We note that for values of x between x_1 and x_2, there are three values of y for each value of x. For values of x between b and x_3 there are two values of y for each value of x. We assume that y is positive between P and T.

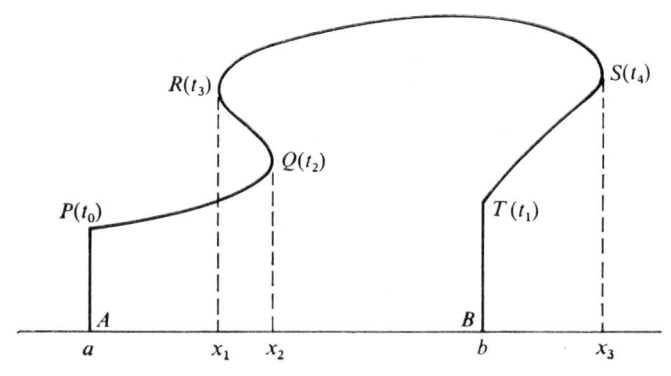

Fig. 7.15

DOUBLE INTEGRALS

Denote the value of y as a point moves along the curve from P to Q by y_1, from Q to R by y_2, from R to S by y_3, and from S to T by y_4; y_1, y_2, y_3, y_4 are single-valued functions of x. Clearly, the required area is

$$A = \int_a^{x_2} y_1\, dx - \int_{x_1}^{x_2} y_2\, dx + \int_{x_1}^{x_3} y_3\, dx - \int_b^{x_3} y_4\, dx.$$

Now suppose that we can find† a parameter t such that x and y are expressed as single-valued differentiable functions of t, $x = f(t)$, $y = \phi(t)$, and that as a point (x, y) moves round the curve from P to T, the parameter t steadily increases or steadily decreases from t_0 to t_1.

Let t_2, t_3, t_4 be the values of the parameter t at the points Q, R, S on the curve. Then

$$A = \int_{t_0}^{t_2} y_1 \frac{dx}{dt} dt - \int_{t_3}^{t_2} y_2 \frac{dx}{dt} dt + \int_{t_3}^{t_4} y_3 \frac{dx}{dt} dt - \int_{t_1}^{t_4} y_4 \frac{dx}{dt} dt.$$

Now, although y_1, y_2, y_3, y_4 are different functions of x, they are the same function of t. y_1 is the value of $\phi(t)$ when t goes from t_0 to t_2; y_2 is the value of $\phi(t)$ when t goes from t_2 to t_3, and so on. Hence, in these four integrals we can simply write $\phi(t)$ or y in place of y_1, y_2, y_3, y_4.

Hence

$$A = \int_{t_0}^{t_2} y \frac{dx}{dt} dt - \int_{t_3}^{t_2} y \frac{dx}{dt} dt + \int_{t_3}^{t_4} y \frac{dx}{dt} dt - \int_{t_1}^{t_4} y \frac{dx}{dt} dt$$

$$= \int_{t_0}^{t_1} y \frac{dx}{dt} dt.$$

If y is negative over part of the range, the results depend upon the particular configuration.

Let

$$I = \int_{t_0}^{t_1} y \frac{dx}{dt} dt.$$

---

† It is not possible to find such a parameter in the case of all curves; those curves for which it is possible are called unicursal.

13

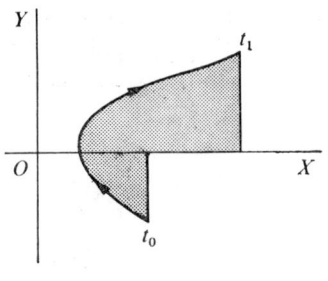

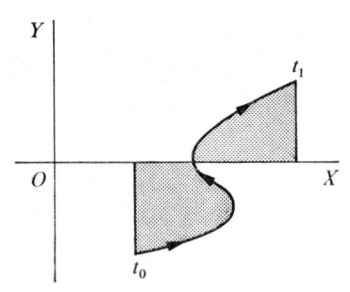

FIG. 7.16   FIG. 7.17

Then for the curve shown in Fig. 7.16, prove that I represents the shaded area; for the curve shown in Fig. 7.17, prove that I represents the difference of the shaded area above the x-axis and the area below the x-axis.

Example. Consider the shaded area (see Fig. 7.18) bounded by the arc PQ of the parabola $y^2 = 4x$, where P is $(1, -2)$ and Q is $(4, 4)$, and the x-axis. The area is

$$A = -\int_0^1 y_1 \, dx + \int_0^4 y_2 \, dx,$$

where
$$y_1 = -2\sqrt{x}, \qquad y_2 = 2\sqrt{x}.$$

Thus
$$A = \int_0^1 2\sqrt{x} \, dx + \int_0^4 2\sqrt{x} \, dx = 12.$$

Alternatively, using the parametric representation $x = t^2$, $y = 2t$ we have

$$A = \int_{-1}^2 y \frac{dx}{dt} dt = \int_{-1}^2 (2t)(2t) \, dt = 12.$$

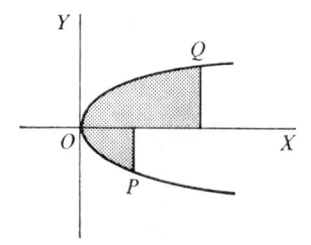

FIG. 7.18

Line integrals

PQ is an arc of a curve with parametric representation $x = f(t)$, $y = \phi(t)$, such that as t steadily increases or steadily decreases from t_0 to t_1 the point (x, y) moves along the curve from P to Q.

Then we shall define the *line integral* $\int_{PQ} f(x, y) \, dx$, where it is understood that the path of integration is along the arc PQ, as

$$\int_{t_0}^{t_1} f(x, y) \frac{dx}{dt} dt.$$

If $f(x, y)$ and $\phi(x, y)$ are functions of x and y, we shall define the line integral

$$\int_{PQ} f(x, y) \, dx + \phi(x, y) \, dy$$

as

$$\int_{t_0}^{t_1} \left\{ f(x, y) \frac{dx}{dt} + \phi(x, y) \frac{dy}{dt} \right\} dt.$$

For example, the line integral $\int x^2 y \, dx + xy^3 \, dy$ taken along the semicircle $x^2 + y^2 = 1$ described positively from $(1, 0)$ to $(-1, 0)$ is equal to

$$\int_0^{\pi} \{(\cos t)^2 (\sin t)(-\sin t) + (\cos t)(\sin t)^3 (\cos t)\} \, dt = -\tfrac{1}{8}\pi + \tfrac{4}{15}.$$

Clearly, the line integral

$$\int_{PQ} f(x, y) \, dx + \phi(x, y) \, dy = - \int_{QP} f(x, y) \, dx + \phi(x, y) \, dy.$$

If y is a single-valued function of x, $y = g(x)$, then clearly we may take x as the parameter t and evaluate the line integral

$$\int_{PQ} f(x, y) \, dx,$$

where at P, $x = a$ and at Q, $x = b$, as

$$\int_a^b f(x, g(x)) \, dx.$$

Also

$$\int_{PQ} \phi(x,y)\,dy = \int_a^b \phi(x, g(x))g'(x)\,dx.$$

Example. Evaluate the line integral $\int xy\,dx$ taken round the triangle vertices $A(0, 0)$, $B(2, 1)$, and $C(1, 2)$, described positively. Along AB, $y = \tfrac{1}{2}x$; along BC, $y = -x+3$; along CA, $y = 2x$. Hence

$$\int_{ABC} xy\,dx = \int_0^2 x(\tfrac{1}{2}x)\,dx + \int_2^1 x(-x+3)\,dx + \int_1^0 x(2x)\,dx$$
$$= -\tfrac{3}{2}.$$

Exercises 7(b)

1. Evaluate the line integral $\int_C xy\,dx$, where C is the semicircle in the first quadrant through $(0, 0)$ and $(4, 0)$ described positively.

2. Evaluate the line integral $\int x^2\,dy$ taken (a) round the circle $x^2+y^2 = 1$ described positively, (b) round the quadrant of the circle from $(-1, 0)$ to $(0, -1)$.

3. Find the value of the line integral $\int xy^2\,dx + (x+y)\,dy$ taken round the triangle vertices $(1, 1)$, $(3, 3)$, $(2, 4)$ described positively.

4. Find the value of the line integral

$$\int_C \left\{ y^2 \sin\left(\frac{\pi x}{4}\right) dx + 2xy \cos\left(\frac{\pi x}{4}\right) dy \right\},$$

where C is the arc of the parabola $y^2 = 4x$ from $(1, 2)$ to $(4, 4)$.

5. Evaluate the line integral $\int_C (x+y)\,dx + (y-x)\,dy$ where C is a path consisting of a straight line from $(0, 1)$ to $(2, 2)$, then the parabolic path $x = 2t$, $y = t^2+1$ from $(2, 2)$ to $(6, 10)$.

6. Evaluate the line integral $\int xy^2\,dx + bx^2\,dy$ taken along the arc of the ellipse $x = a\cos t$, $y = b\sin t$ from $(a, 0)$ to $(0, b)$.

7. Evaluate the line integral $\int xy\,dx + (x^2+y^2)\,dy$ taken along the astroid $x = a\cos^3 t$, $y = a\sin^3 t$ from $(0, a)$ to $(a, 0)$.

8. Evaluate the line integral $\displaystyle\int \frac{x^2y\,dx + x^2\,dy}{x^2+y^2}$ taken positively round the square vertices $(\pm 1, \pm 1)$.

The planimeter

An instrument for measuring the area of a plane closed curve is the planimeter, invented by Amsler in 1854. It consists essentially of a bar OA pivoted at a fixed point O, and a second bar AP, pivoted at A (see Fig. 7.19). The point P is made to move in an anticlockwise direction round the closed curve C whose area we wish to find. A small wheel is fixed at P, the plane of the wheel being perpendicular to AP, and a revolution counter on the wheel measures the total displacement of P perpendicular to AP as P describes the curve.

Let $OA = a$, $AP = b$, the angle made by OA with a fixed line OX in the plane be θ, and the angle made by AP with OX be ϕ.

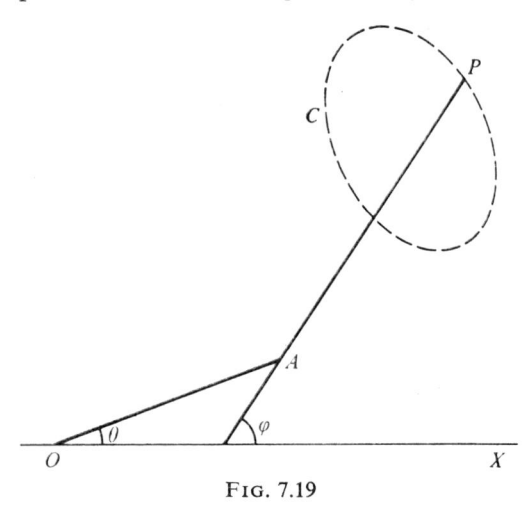

FIG. 7.19

Then in a small displacement of P, the component of displacement perpendicular to AP is

$$b\delta\phi + (a\delta\theta)\cos(\phi-\theta),$$

and the total component displacement of P as P describes the curve is

$$\int_C \{b\,d\phi + a\cos(\phi-\theta)\,d\theta\} = \int_C a\cos(\phi-\theta)\,d\theta,$$

since $\int_C d\phi = 0$ if A lies outside C. Now the coordinates of a point on the curve are given by

$$x = a\cos\theta + b\cos\phi,$$
$$y = a\sin\theta + b\sin\phi.$$

Using the result for the area of the closed curve,

$$A = \frac{1}{2}\int_C \left(x\frac{dy}{dt} - y\frac{dx}{dt}\right)dt,$$

we see that

$$A = \frac{1}{2}\int_C (a\cos\theta + b\cos\phi)(a\cos\theta\, d\theta + b\cos\phi\, d\phi) -$$

$$-(a\sin\theta + b\sin\phi)(-a\sin\theta\, d\theta - b\sin\phi\, d\phi)$$

$$= \frac{1}{2}\int_C a^2\, d\theta + b^2\, d\phi + ab\cos(\phi-\theta)(d\theta+d\phi).$$

Now, if O is a point lying outside the curve C, clearly

$$\int_C d\theta = 0; \quad \text{also} \quad \int_C d\phi = 0.$$

Hence

$$A = \tfrac{1}{2}ab\int_C \cos(\phi-\theta)(d\theta+d\phi)$$

$$= \tfrac{1}{2}ab\int_C \cos(\phi-\theta)\{d(\phi-\theta)+2\,d\theta\}$$

$$= \tfrac{1}{2}ab[\sin(\phi-\theta)]_C + ab\int_C \cos(\phi-\theta)\,d\theta$$

$$= ab\int_C \cos(\phi-\theta)\,d\theta.$$

Now we have shown that the total component displacement of P perpendicular to OP is

$$a\int_C \cos(\phi-\theta)\,d\theta,$$

and hence the area of the curve is equal to b times this displacement. Thus from the measure of the displacement shown on the revolution counter we can calculate the area of the curve.

Double integrals

Let C be a simple closed curve in the XOY plane, and denote by A the domain bounded by C (see Fig. 7.20); $f(x, y)$ is a given

DOUBLE INTEGRALS

function of x and y. We divide the domain A into small regions by a network of lines parallel to the coordinate axes (see Fig. 7.21); a typical element of the mesh is a rectangle bounded by the lines

$$x = x_r, \quad x = x_r + \delta x_r; \quad y = y_s, \quad y = y_s + \delta y_s.$$

Let $m_{r,s}$ denote the minimum value and $M_{r,s}$ the maximum value of $f(x, y)$ in this rectangle. We now form the sums

$$\sum_r \sum_s m_{r,s}\, \delta x_r\, \delta y_s,$$

and

$$\sum_r \sum_s M_{r,s}\, \delta x_r\, \delta y_s,$$

the summations being taken over all the rectangular elements of the mesh. If each of these sums tends to a limit as the sides of the

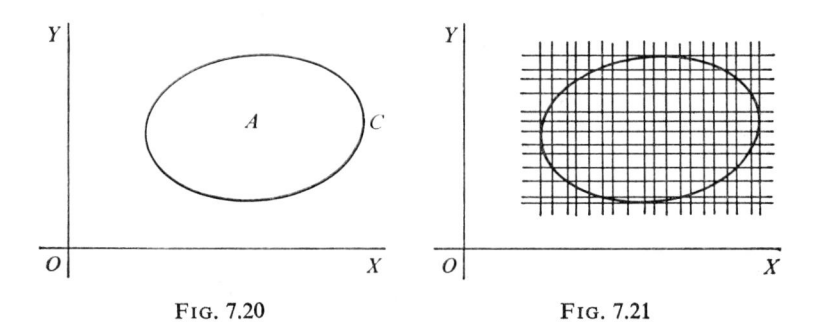

FIG. 7.20   FIG. 7.21

elements of the mesh tend to zero, i.e. as δx and δy tend to zero, and the two limits are the same, we define the double integral

$$\iint_A f(x, y)\, dx\, dy$$

as the value of the common limit. It may be proved that the two sums have a common limit if $f(x, y)$ is a continuous function of x and y over the domain A, but this is not the only type of function for which the common limit exists.

Since $m_{r,s} \leq f(x_r, y_s) \leq M_{r,s}$, it is clear that

$$\sum_r \sum_s f(x_r, y_s)\, \delta x_r\, \delta y_s$$

will also tend to this common limit and hence we may write

$$\iint_A f(x, y)\, dx\, dy = \lim_{\substack{\delta x \to 0 \\ \delta y \to 0}} \sum \sum f(x, y)\, \delta x\, \delta y.$$

Evaluation of double integrals as repeated integrals

In evaluating the sum $\Sigma \Sigma f(x, y)\delta x \delta y$ we can take the elementary rectangles in any desired pattern, but two patterns are specially useful, by strips parallel to OY, and secondly by strips parallel to OX.

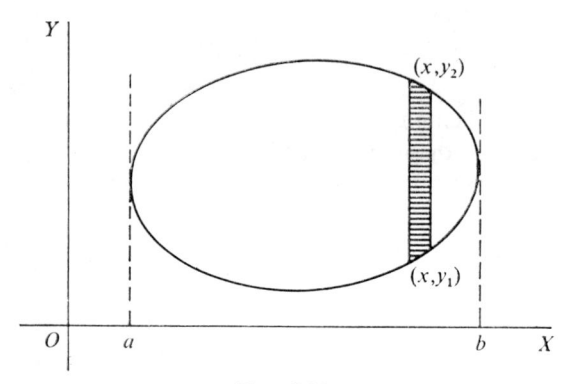

FIG. 7.22

Let us keep x and δx fixed and sum $f(x, y)\delta x \delta y$ along the strip parallel to OY giving $\delta x \Sigma f(x, y)\delta y$ (see Fig. 7.22). The ordinate at x cuts the curve in two points (x, y_1) and (x, y_2), where y_1 and y_2 are functions of x. As δy tends to 0, the sum $\delta x \Sigma f(x, y)\delta y$ tends to the limit

$$\delta x \int_{y_1}^{y_2} f(x, y)\, dy.$$

This integral, when evaluated, is a certain function of x.

We next sum over all the strips from $x = a$ to $x = b$. This gives

$$\sum \delta x \left\{ \int_{y_1}^{y_2} f(x, y)\, dy \right\}.$$

As δx tends to zero, this sum tends to the limit

$$\int_a^b dx \left\{ \int_{y_1}^{y_2} f(x, y)\, dy \right\}.$$

Thus, we can represent the double integral $\iint_A f(x, y)\, dx\, dy$ in the form

$$\int_a^b dx \left\{ \int_{y_1}^{y_2} f(x, y)\, dy \right\},$$

called a repeated integral.

Alternatively, we can evaluate the sum $\Sigma \Sigma f(x, y) \delta x \delta y$ by strips parallel to OX. Let us keep y and δy fixed and sum $f(x, y)\delta x \delta y$ along the strip parallel to OX, giving $\delta y \Sigma f(x, y)\delta x$ (see Fig. 7.23).

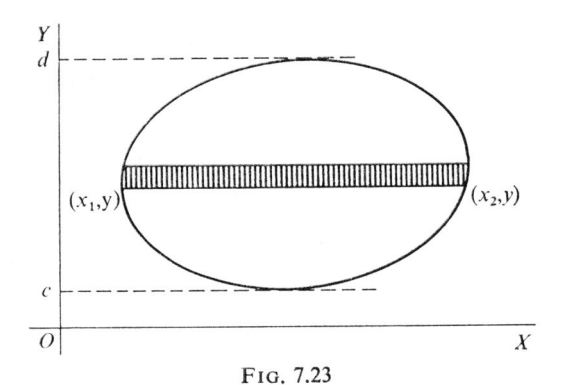

Fig. 7.23

The abscissa at y cuts the curve in two points (x_1, y) and (x_2, y), where x_1 and x_2 are functions of y. As δx tends to 0, the sum $\delta y \Sigma f(x, y)\delta x$ tends to the limit

$$\delta y \int_{x_1}^{x_2} f(x, y)\, dx.$$

The integral, when evaluated, is a certain function of y. We next sum over all the strips from $y = c$ to $y = d$. This gives

$$\sum \delta y \left\{ \int_{x_1}^{x_2} f(x, y)\, dx \right\}.$$

As δy tends to 0, this sum tends to the limit

$$\int_c^d dy \left\{ \int_{x_1}^{x_2} f(x, y)\, dx \right\}.$$

Thus, we can represent the double integral $\iint_A f(x, y)\, dx\, dy$ in the form

$$\int_c^d dy \left\{ \int_{x_1}^{x_2} f(x, y)\, dx \right\},$$

another repeated integral. We have thus shown that

$$\iint_A f(x, y)\, dx\, dy = \int_a^b dx \int_{y_1}^{y_2} f(x, y)\, dy = \int_c^d dy \int_{x_1}^{x_2} f(x, y)\, dx.$$

Example. Evaluate $I = \iint_A x^2 y\, dx\, dy$, where A is the domain bounded by $x = 0$, $y = 0$, $x+y = 1$ (see Fig. 7.24). For a given value of x,

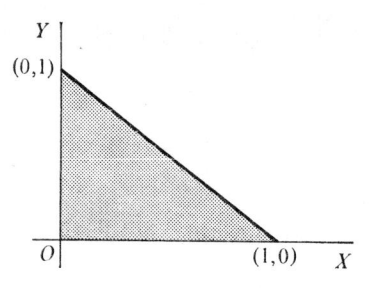

FIG. 7.24

the range of values of y is from 0 to $1-x$. Hence

$$I = \int_0^1 dx \int_0^{1-x} x^2 y\, dy$$

$$= \int_0^1 dx \left[\frac{x^2 y^2}{2} \right]_0^{1-x}$$

$$= \int_0^1 dx \frac{x^2(1-x)^2}{2} = \frac{1}{60}.$$

Alternatively, for a given value of y, the range of values of x is from

0 to $1-y$. Hence

$$I = \int_0^1 dy \int_0^{1-y} x^2 y \, dx$$

$$= \int_0^1 dy \left[\frac{x^3 y}{3} \right]_0^{1-y}$$

$$= \int_0^1 dy \frac{(1-y)^3 y}{3} = \frac{1}{60}.$$

Example A is the domain in the first quadrant bounded by the circle $x^2 + y^2 = 25$ and the hyperbola $xy = 12$. Evaluate

$$I = \iint_A x(1+y) \, dx \, dy.$$

The domain is shown in Fig. 7.25. For a given value of x between 3 and 4, y ranges from

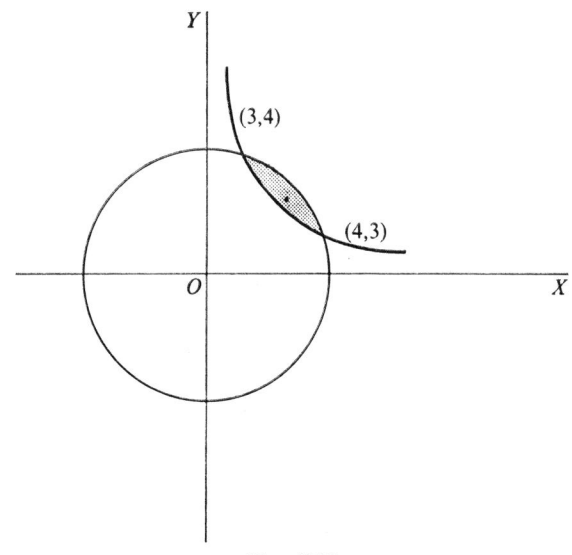

$$\frac{12}{x} \quad \text{to} \quad \sqrt{(25-x^2)}.$$

Fig. 7.25

Hence
$$I = \int_3^4 dx \int_{12/x}^{\sqrt{(25-x^2)}} x(1+y)\,dy$$
$$= \int_3^4 dx\, x \left[y + \frac{y^2}{2}\right]_{12/x}^{\sqrt{(25-x^2)}}$$
$$= \int_3^4 dx\, x \left[\sqrt{(25-x^2)} + \frac{25-x^2}{2} - \frac{12}{x} - \frac{72}{x^2}\right]$$
$$= \left[-\tfrac{1}{3}(25-x^2)^{\frac{3}{2}} + \tfrac{25}{4}x^2 - \tfrac{1}{8}x^4 - 12x - 72 \log x\right]_3^4$$
$$= \tfrac{533}{24} - 72 \log \tfrac{4}{3}.$$

Alternatively, for a given value of y between 3 and 4, x ranges from $\dfrac{12}{y}$ to $\sqrt{(25-y^2)}$ and hence

$$I = \int_3^4 dy \int_{12/y}^{\sqrt{(25-y^2)}} x(1+y)\,dx$$
$$= \int_3^4 dy(1+y) \left[\frac{x^2}{2}\right]_{12/y}^{\sqrt{(25-y^2)}}$$
$$= \int_3^4 dy(1+y)\left(\frac{25-y^2}{2} - \frac{72}{y^2}\right) = \frac{533}{24} - 72 \log \frac{4}{3}.$$

Conditions for the existence of a double integral

We have already stated that a sufficient condition for the existence of the double integral $\iint_A f(x, y)\, dx\, dy$ is that $f(x, y)$ is a continuous function of x and y over the domain A.

The necessity for investigation of the behaviour of $f(x, y)$ over the domain of integration is illustrated by the following example.

Example. Prove that the value of the repeated integral

$$\int_0^1 dx \int_0^1 \frac{x-y}{(x+y)^3}\, dy = \tfrac{1}{2},$$

and the value of the repeated integral

$$\int_0^1 dy \int_0^1 \frac{x-y}{(x+y)^3} dx = -\tfrac{1}{2}.$$

The fact that these answers are different shows that the double integral in the evaluation of which these repeated integrals would occur, namely,

$$\iint_A \frac{x-y}{(x+y)^3} dx\, dy,$$

where A denotes the square domain with vertices $(0, 0)$, $(1, 0)$, $(1, 1)$, $(0, 1)$, cannot exist. The reason for the non-existence of the double integral is clear for the function $\frac{x-y}{(x+y)^3}$ is not defined at the point $(0, 0)$ of the domain.

Volumes of solids

Consider the solid with plane base A bounded by a simple curve C in the xy-plane, a cylinder through C with generators parallel to OZ, and the surface $z = f(x, y)$ where $z \geq 0$ for all points (x, y) of A. Clearly, the volume of the solid is

$$\lim_{\substack{\delta x \to 0 \\ \delta y \to 0}} \sum \sum z\, \delta x\, \delta y = \iint_A z\, dx\, dy = \iint_A f(x, y)\, dx\, dy.$$

Example. Find the volume bounded by the surface $z = x^2 + 6y^2$, the cylinder $x^2 + 4y^2 = 4$ and the plane $z = 0$.

$$V = \iint z\, dx\, dy = \iint (x^2 + 6y^2)\, dx\, dy$$

$$= \int_{-2}^{2} dx \int_{-\tfrac{1}{2}\sqrt{(4-x^2)}}^{\tfrac{1}{2}\sqrt{(4-x^2)}} (x^2 + 6y^2)\, dy = \int_{-2}^{2} dx [x^2 y + 2y^3]_{-\tfrac{1}{2}\sqrt{(4-x^2)}}^{\tfrac{1}{2}\sqrt{(4-x^2)}}$$

$$= \int_{-2}^{2} (2 + \tfrac{1}{2}x^2)\sqrt{(4-x^2)}\, dx = 5\pi.$$

Exercises 7(c)

1. Evaluate $\iint_A xy(x+y)\, dx\, dy$, where A is the rectangle bounded by the lines $x = 2, 4; y = 1, 3$.

2. Evaluate $\iint_A xy^3 \, dx \, dy$, where A is the domain bounded by the parabola $y^2 = 4x$ and the lines $x = 1$, $y = 1$.

3. Find the volume of the solid on a square base with sides $x = a, 3a$; $y = 2a, 4a$, faces perpendicular to the base, and bounded opposite to the base by the surface $4az = x^2 + xy + y^2$.

4. An area is bounded by $x^2 = 4y$, $x = 0$, $y = 1$. Through points on the boundary of the area, lines are drawn perpendicular to the plane XOY to intersect the surface $z = x^2y$. Find the volume of the solid so formed.

5. Prove that
$$\int_0^a dx \int_0^x f(x, y) \, dy = \int_0^a dy \int_y^a f(x, y) \, dx.$$

6. A lamina in the XOY plane is such that the density ρ of the lamina at any point (x, y) on it is a function of x and y. Show that the coordinates of the centre of mass of the lamina are given by
$$\bar{x} = \frac{\iint \rho x \, dx \, dy}{\iint \rho \, dx \, dy}, \quad \bar{y} = \frac{\iint \rho y \, dx \, dy}{\iint \rho \, dx \, dy}.$$

A circular disc, centre O radius a, has variable density, the density at any point (x, y) being equal to $k(x^2 + y^2)x^2$, where k is a constant. Find the mass of the disc in terms of a and k.

Find also the position of the centre of mass of the half of the disc for which x is positive.

7. Prove that the moment of inertia of a lamina in the XOY plane about the axis OX is equal to
$$\iint \rho y^2 \, dx \, dy,$$
and find an expression for the moment of inertia about OY.

If the lamina is a triangle bounded by the line $x + y = 1$ and the co-ordinate axes, and $\rho = (x + y)^2$, find the moment of inertia of the lamina about OX.

8. Prove that the volume of the wedge intercepted between the cylinder $x^2 + y^2 = 2x$ and the planes $z = 0$, $z = x$ is π.

9. Evaluate the double integral $\iint z \, dx \, dy$, where $z = \dfrac{x^2}{a^2} + \dfrac{y^2}{b^2}$ and the region of integration is the area
$$\frac{x^2}{a^2} + \frac{y^2}{b^2} \leq 1.$$

10. ABC is a triangle in the plane XOY bounded by the lines $x = 2$, $y = 1$, $y = x$. A solid is formed with ABC as its base, planes through

DOUBLE INTEGRALS

AB, BC, CA perpendicular to XOY as its sides and its upper surface given by $z = xy + y^2$. Express the volume of the solid as a double integral, and evaluate this integral in two ways by repeated integrals. Hence show that the volume of the solid is $\frac{49}{24}$.

11. Find $\iint x^2 y^3 \, dx \, dy$ over the triangle defined by $x \geq 0, y \geq 0, x + 2y \leq 2$.

12. Find the volume bounded by the cylinder $(x-2)^2 + (y-2)^2 = 1$, the plane $z = 0$, and the surface $z = xy$, lying wholly in the first octant.

13. The cross section of a cylinder is an ellipse of semi-axes a and b. Find the volume cut off by a plane through the minor axis of the base of the cylinder making angle θ with the plane of the ellipse.

14. Change the order of integration in the repeated integrals

$$\int_0^a dx \int_{b/(a+b)}^{b/(x+b)} f(x, y) \, dy \quad (a > 0, b > 0), \quad \int_1^2 dx \int_0^{1/x} f(x, y) \, dy.$$

15. D is the smaller of the two regions bounded by the ellipse $x^2 + 4y^2 = 2$ and the parabola $4y = x^2 + 1$. Evaluate the integral

$$\iint_D (x^2 + y) \, dx \, dy.$$

16. If $f(x, y)$ is a continuous function of x and y over a certain triangular region, change the order of integration in the repeated integral

$$\int_{\pi/2}^{\pi} dx \int_0^{x-(\pi/2)} f(x, y) \, dy.$$

Check your answer by considering the values of the two integrals when $f(x, y) = (x + y) \sin x$.

17. If A denotes the interior of the ellipse with equation

$$x^2 + 4y^2 - 2x + 8y + 1 = 0,$$

evaluate the integral $\iint_A (y^3 + y + 1) \, dx \, dy$.

18. Evaluate the line integrals $\int_{PQ} xy \, dx + (x^2 + y^2) \, dy$, where P is the point $(1, 2)$, and Q is the point $(4, 8)$ along the following paths of integration:
    (i) a straight line path from P to Q;
    (ii) a straight line path from P to R $(3, 3)$, followed by a straight line path from R to Q;
    (iii) the parabola $x = 2t + 2, y = 8t^2$.

19. A particle mass m moves along the path PQ in the xy-plane in a field of force whose components at any point (x, y) are X_1 parallel to OX and Y_1 parallel to OY. Express as a line integral the work done against the forces of the field in moving the particle from P to Q.

If $X_1 = x+2y$ and $Y_1 = 2x+y$, calculate the work done in moving the particle along the x-axis from $P(-2a, 0)$ to $Q(2a, 0)$ and then round the semi-ellipse $x = 2a \cos t$, $y = a \sin t$ in the upper half of the xy-plane from Q to P.

20. A is the domain bounded by the curves $xy = 1$, $x^3y = 1$ and the line $x = 2$. Prove that $\iint_A (x+y)^2 \, dx \, dy = \frac{2763}{2048}$.

21. A is the parallelogram with vertices $(0, 1)$, $(2, 2)$, $(0, -1)$, $(-2, -2)$. Evaluate the double integral $\iint_A x(x+y) \, dx \, dy$.

Surface integrals

We shall now obtain an integral for the area of the portion of the surface $z = f(x, y)$ cut off by a cylinder whose base is a simple closed curve C in the xy-plane and having generators parallel to OZ.

The element of the surface cut off by a prism base $\delta x \delta y$ is $\delta x \delta y / \cos \theta$, where θ is the angle between the normal to the surface and OZ (see Fig. 7.26). Now, the tangent plane at any point (x_1, y_1, z_1) of the surface has equation

$$(x-x_1)\frac{\partial f}{\partial x_1} + (y-y_1)\frac{\partial f}{\partial y_1} + (z-z_1)(-1) = 0,$$

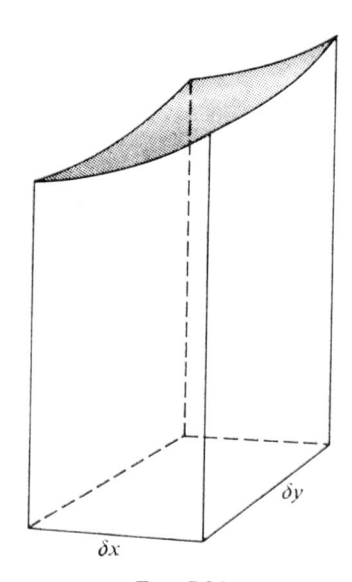

FIG. 7.26

DOUBLE INTEGRALS

from which it follows that the direction cosines of the normal are proportional to $\dfrac{\partial f}{\partial x_1}, \dfrac{\partial f}{\partial y_1}, -1$. Hence

$$\cos\theta = \dfrac{1}{\sqrt{\left\{\left(\dfrac{\partial f}{\partial x}\right)^2 + \left(\dfrac{\partial f}{\partial y}\right)^2 + 1\right\}}},$$

and thus the element of surface is

$$\sqrt{\left\{\left(\dfrac{\partial f}{\partial x}\right)^2 + \left(\dfrac{\partial f}{\partial y}\right)^2 + 1\right\}} \, \delta x \, \delta y.$$

Hence, the area of the surface is given by the double integral

$$\iint_A \sqrt{\left\{\left(\dfrac{\partial f}{\partial x}\right)^2 + \left(\dfrac{\partial f}{\partial y}\right)^2 + 1\right\}} \, dx \, dy,$$

where A is the domain bounded by C.

If the equation of the surface is given in the form $f(x, y, z) = 0$, it is easily seen that the portion of the surface cut off by the cylinder has area

$$\iint_A \dfrac{\sqrt{\left\{\left(\dfrac{\partial f}{\partial x}\right)^2 + \left(\dfrac{\partial f}{\partial y}\right)^2 + \left(\dfrac{\partial f}{\partial z}\right)^2\right\}}}{\left|\dfrac{\partial f}{\partial z}\right|} \, dx \, dy.$$

We shall see later that some of the rather complicated integrals we obtain for surface area can be simplified by changing the variables in the double integral.

Example. Find the area of the surface $z = \sqrt{(x^2+y^2)}$ cut off by the cylinder $x^2+y^2 = r^2$. In this case

$$\sqrt{\left\{1+\left(\dfrac{\partial z}{\partial x}\right)^2 + \left(\dfrac{\partial z}{\partial y}\right)^2\right\}} = \sqrt{\left\{1 + \dfrac{x^2}{x^2+y^2} + \dfrac{y^2}{x^2+y^2}\right\}} = \sqrt{2}.$$

Hence the required surface area is

$$S = \iint_A \sqrt{2} \, dx \, dy = \sqrt{2}(\pi r^2).$$

This is a well-known result since the surface $z = \sqrt{(x^2+y^2)}$ is a cone, vertex 0, and we are simply finding the curved surface area of a cone with radius of base r and height r.

Exercises 7(d)

1. Find the area of the surface of the hemisphere $x^2+y^2+z^2 = a^2$, $z \geq 0$, cut off by the cylinder $x^2+y^2-ax = 0$.

2. Prove that the area of the surface of the cone $z^2 = 2(x^2+y^2)$ cut off by the cylinder $y^2 = x$, $x \leq 1$, is $4/\sqrt{3}$.

3. Investigate the intersection of the octant of the ellipsoid

$$x^2+4y^2+2z^2 = 1,$$

$x \geq 0$, $y \geq 0$, $z \geq 0$, with the surface $z^2 = 2xy$.

Prove that the area of that part of the surface $z^2 = 2xy$ which lies within the octant of the ellipsoid can be expressed in the form of the double integral

$$\iint \frac{x+y}{\sqrt{(2xy)}} \, dx \, dy$$

taken over the triangular domain defined by $x \geq 0$, $y \geq 0$, $x+2y \leq 1$.

Hence show that the surface area is $\dfrac{3\pi}{16}$.

Green's theorem

Denote by C a simple closed curve bounding a domain A (see Fig. 7.27). Let $P(x, y)$ and $Q(x, y)$ denote two functions of x and y, defined within and on the contour C, and continuous with continuous partial derivatives in the region.

Let a line parallel to OX meet the curve in points (x_1, y) and

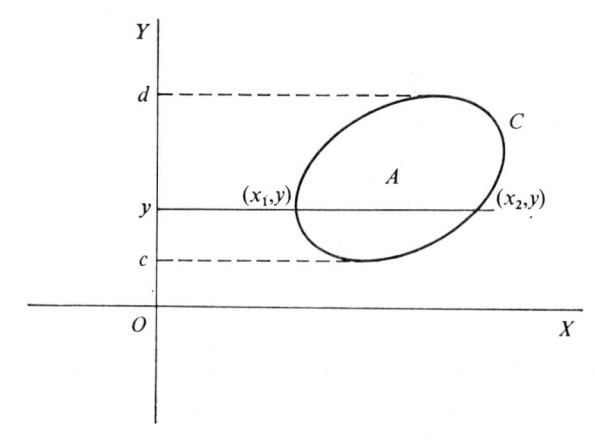

FIG. 7.27

(x_2, y); x_1 and x_2 will be functions of y. Then

$$\iint_A \frac{\partial Q}{\partial x}\, dx\, dy = \int_c^d dy \int_{x_1}^{x_2} \frac{\partial Q}{\partial x}\, dx$$

$$= \int_c^d dy [Q]_{x_1}^{x_2} = \int_c^d dy \{Q(x_2, y) - Q(x_1, y)\}$$

$$= \int_C Q\, dy,$$

the last integral being taken round the curve C, described in a positive direction.

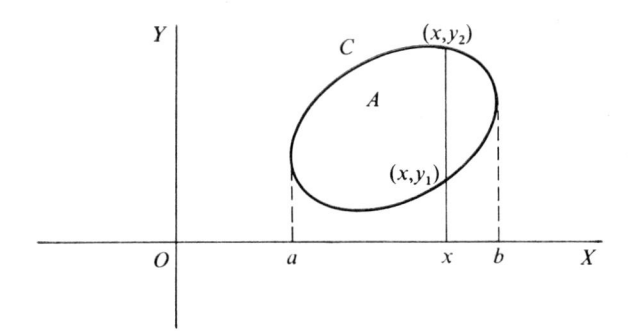

Fig. 7.28

Next, (see Fig. 7.28) let a line parallel to OY meet the curve in points (x, y_1) and (x, y_2); y_1 and y_2 will be functions of x. Then

$$\iint_A \frac{\partial P}{\partial y}\, dx\, dy = \int_a^b dx \int_{y_1}^{y_2} \frac{\partial P}{\partial y}\, dy$$

$$= \int_a^b dx [P]_{y_1}^{y_2} = \int_a^b dx \{P(x, y_2) - P(x, y_1)\}$$

$$= -\int_C P\, dx,$$

the last integral being taken round the curve C, described in a positive direction.

It follows that

$$\iint_A \left(\frac{\partial Q}{\partial x} - \frac{\partial P}{\partial y}\right) dx\, dy = \int_C P\, dx + Q\, dy.$$

This result is known as Green's theorem and has an important role in many branches of mathematics, including the theory of functions of a complex variable and aerodynamics.

We shall now give a verification of Green's theorem in a simple case. Take as the contour C the circle

$$(x-1)^2 + (y-1)^2 = 1,$$

and let $P = xy$, $Q = x^2 + y^2$. Then

$$\iint_A \left(\frac{\partial Q}{\partial x} - \frac{\partial P}{\partial y}\right) dx\, dy = \iint_A (2x - x)\, dx\, dy$$

$$= \int_0^2 x\, dx \int_{1-\sqrt{1-(x-1)^2}}^{1+\sqrt{1-(x-1)^2}} dy = \int_0^2 2x\sqrt{(1-(x-1)^2)}\, dx,$$

which gives, on writing $x - 1 = \sin\theta$,

$$\int_{-\pi/2}^{\pi/2} 2(1+\sin\theta)\cos\theta \cdot \cos\theta\, d\theta = \pi.$$

$\int_C P\, dx + Q\, dy$ can be evaluated directly by expressing x and y in the form $x = 1 + \cos t$, $y = 1 + \sin t$, and the integral as

$$\int_0^{2\pi} \left(P\frac{dx}{dt} + Q\frac{dy}{dt}\right) dt,$$

which then becomes

$$= \int_0^{2\pi} \{(1+\cos t)(1+\sin t)(-\sin t) + (3 + 2\sin t + 2\cos t)(\cos t)\}\, dt$$

$$= \pi.$$

To illustrate the importance of the conditions of validity of Green's theorem, consider the value of the line integral

$$I = \int_C \frac{x^2 y\, dx - x^3\, dy}{(x^2+y^2)^2},$$

where C is the circle $x^2+y^2 = a^2$, $0 < a < 1$. On writing $x = a\cos\theta$, $y = a\sin\theta$, we have

$$I = \int_0^{2\pi} \frac{(a\cos\theta)^2(a\sin\theta)(-a\sin\theta)\, d\theta - (a\cos\theta)^3 a\cos\theta\, d\theta}{a^4}$$

$$= \int_0^{2\pi} (-2\sin^2\theta \cos^2\theta - \cos^4\theta)\, d\theta = -\pi.$$

It is *wrong* to attempt to use Green's theorem and express the line integral as

$$\iint_A \left\{ \frac{\partial}{\partial x}\left(\frac{-x^3}{(x^2+y^2)^2}\right) - \frac{\partial}{\partial y}\left(\frac{x^2 y}{(x^2+y^2)^2}\right) \right\} dx\, dy$$

$$= \iint_A \left\{ \frac{x^2(x^2-3y^2)}{(x^2+y^2)^3} - \frac{x^2(x^2-3y^2)}{(x^2+y^2)^3} \right\} dx\, dy = 0,$$

because $\dfrac{x^2 y}{(x^2+y^2)^2}$ and $\dfrac{x^3}{(x^2+y^2)^2}$ do not have continuous first derivatives in the closed region $x^2+y^2 \le a^2$; in fact, the functions do not exist at the origin O. But, in calculating the line integral

$$I = \int_C \frac{x^2 y\, dx - x^3\, dy}{(x^2+y^2)^2},$$

where C is the circle $(x-a)^2+(y-2a)^2 = a^2$, Green's theorem may be used.

Exercises 7(e)

1. Verify Green's theorem for

$$P = xy + y^2, \qquad Q = x^2 - y^2,$$

where the contour C is the closed curve bounded by the parabolas $y^2 = x$ and $x^2 = y$.

2. By expressing the line integral $\int_C x\,dy - y\,dx$ as a double integral using Green's theorem, establish the well-known result for the area of a simple closed curve

$$A = \frac{1}{2}\int_C x\,dy - y\,dx.$$

3. The closed curve C consists of the arc of the parabola $x^2 = 4ay$ between the points $P(2a, a)$ and $Q(-2a, a)$ and the straight line QP. The region enclosed by C is A.

The integral I is defined by

$$I = \int_C xy^2\,dx - x^2y\,dy,$$

where C is described positively. Express I as a double integral J using Green's theorem and evaluate I and J independently to verify Green's theorem.

4. A is the point $(1, 1)$, B is the point $(3, 2)$. Prove that the value of the line integral $\int_{AB} (3x^2y + xy^2 + 6x)\,dx + (x^3 + x^2y - y)\,dy$ is independent of the path joining A and B and evaluate the integral.

5. C is the parallelogram with vertices $(0, 0)$, $(1, 1)$, $(2, 3)$, and $(1, 2)$.

I denotes the line integral $\int_C (x^2 + y)\,dx$. Evaluate I in two ways, (i) directly, (ii) by converting the integral into a double integral using Green's theorem.

6. C denotes the closed curve formed by the straight line joining $P(1, 0)$ to $Q(3, 0)$ and the semicircle in the upper half of the XOY plane with PQ as diameter. I denotes the line integral

$$I = \int_C (x^2 + y^2)\,dx - (xy^3 - 6x^2)\,dy.$$

Evaluate I directly and also by converting it into a double integral.

Change of variables in a double integral

Mapping

Consider the pair of transformations $x = u + v$, $y = u - v$, and the equivalent pair $u = \frac{1}{2}(x + y)$, $v = \frac{1}{2}(x - y)$.

It is clear that corresponding to any point in the uv-plane there is a unique point in the xy-plane; corresponding to any point in the xy-plane there is a unique point in the uv-plane. We say there is a one-to-one correspondence between the xy-plane and the uv-plane; and there is a one-to-one correspondence between any given set of points in the xy-plane and the corresponding set of points in the uv-plane.

If we apply these transformations to points of the square domain $OABC$ in the xy-plane (Fig. 7.29) we obtain a set of points in the uv-plane.

$$OA \to u-v = 0; \quad O \to O_1(0, 0); \quad A(1, 0) \to A_1(\tfrac{1}{2}, \tfrac{1}{2}),$$

and the set of points in the interval OA transforms into the set of points in the interval $O_1 A_1$.

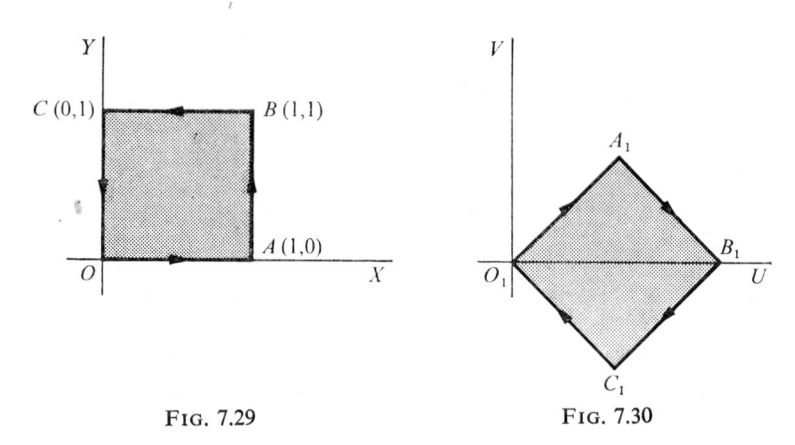

FIG. 7.29   FIG. 7.30

Similarly $AB \to u+v = 1$, $BC \to u-v = 1$, $CO \to u+v = 0$. It will be clear that the square domain $OABC$ in the xy-plane is transformed into a square domain $O_1 A_1 B_1 C_1$ in the uv-plane (Fig. 7.30). If the boundary $OABC$ is described positively, the boundary $O_1 A_1 B_1 C_1$ is described negatively.

In general, the pair of equations

$$u = f(x, y),$$
$$v = \phi(x, y),$$

define a transformation from points in the xy-plane to points in the uv-plane. The pair of equations

$$x = f_1(u, v),$$
$$y = \phi_1(u, v),$$

define a transformation from points in the uv-plane to points in the xy-plane. Each pair of equations establishes a correspondence between points in the xy-plane and points in the uv-plane.

If a region R in the xy-plane is transformed into the region R_1 in the uv-plane we say that the region R is mapped into the region R_1.

The most important type of mapping is a 1–1 mapping in which to each point of the region R in the xy-plane there corresponds a unique point of the region R_1 in the uv-plane and conversely, to each point of the region R_1 there corresponds a unique point of R.

In the case of 1–1 mapping, the pair of equations

$$\left.\begin{array}{l}u = f(x,y)\\ v = \phi(x,y)\end{array}\right\} \text{ can be transformed into } \left.\begin{array}{l}x = f_1(u,v)\\ y = \phi_1(u,v)\end{array}\right\},$$

where f, ϕ, f_1, ϕ_1 are single-valued functions.

For the linear transformations

$$x = au + bv,$$
$$y = cu + dv, \qquad ad - bc \neq 0,$$

there is a 1–1 mapping of the whole of the xy-plane onto the uv-plane, but for other transformations there is normally a limitation on the regions for which the mapping is 1–1.

For example, consider the transformations $x = u^2$, $y = v^2$. Then corresponding to each point in the uv-plane there is a unique point in the xy-plane, but corresponding to each point in the first quadrant of the xy-plane, there are four points in the uv-plane. For example, the point $x = 1, y = 1$ transforms into $u = \pm 1, v = \pm 1$. This is not a 1–1 mapping. The transformation does, however, give a 1–1 mapping of the first quadrant of the xy-plane onto the first quadrant of the uv-plane.

Exercises 7(f)

1. The vertices of a triangle in the xy-plane are $(1, 1)$, $(2, 0)$, $(3, 5)$. The triangle is mapped into the uv-plane by the transformations

$$x = u + 2v,$$
$$y = v + 2u.$$

Find the domain in the uv-plane and state whether the mapping is a 1–1 mapping. If the boundary of the triangle in the xy-plane is described positively, state how the boundary of the domain in the uv-plane is described.

DOUBLE INTEGRALS

2. The rectangle in the xy-plane bounded by the lines $x = 1$, $x = 2$, $y = 0$, $y = 2$ is mapped into the uv-plane by the transformations

$$x = 2u - v,$$
$$y = 3u + 2v.$$

Prove that the rectangle is mapped into a parallelogram in the uv-plane, and that if the boundary of the rectangle is described positively the boundary of the parallelogram is described positively. Prove that the mapping is 1–1.

3. A region R in the xy-plane consists of the ring between the circles $x^2 + y^2 = 1$, $x^2 + y^2 = 4$. Determine the region R_1 into which R is mapped by the transformation $u = x^2$, $v = y^2$. Is this a 1–1 mapping? Answer the same questions for a region R which is the portion of the ring in the first quadrant.

4. A region R in the xy-plane consists of the ring between the circles $x^2 + y^2 = 4$, $x^2 + y^2 = 9$. Determine the region R_1 into which R is mapped by the transformation $x = \sqrt{u}$, $y = \sqrt{v}$. Is this a 1–1 mapping?

5. A region R in the xy-plane is defined by $x + y \leq 1$, $x \geq 0$, $y \geq 0$. Find the region R_1 into which R is mapped under the transformations $x = u^2$, $y = v^2$. Is this a 1–1 mapping?

6. A region R in the xy-plane is defined by $x^2 + y^2 \leq 1$, $x \geq 0$, $y \geq 0$. Determine the region R_1 into which R is mapped under the transformation $x = u \cos v$, $y = u \sin v$, where $u \geq 0$, $0 \leq v < 2\pi$.

7. R is a domain in the xy-plane bounded by a parallelogram with vertices $(1, 1)$, $(2, 2)$, $(4, 5)$, $(3, 4)$. Find the region R_1 into which R is mapped by the transformation

$$x = u + v,$$
$$y = u + 2v.$$

Is this a 1–1 mapping?

8. R is the sector of a circle in the xy-plane defined by

$$x^2 + y^2 \leq 4, \quad x \geq 0, \quad y \geq 0, \quad y \leq x.$$

Find the domain in the uv-plane into which the sector is mapped by the transformation

$$x = \sqrt[4]{u}, \quad y = \sqrt[4]{v}.$$

State whether the mapping is 1–1.

Jacobians

Consider a simple closed curve C in the xy-plane bounding a domain A and let

$$x = f(u, v),$$
$$y = \phi(u, v),$$

be a 1–1 mapping which transforms C into a simple closed curve C_1 in the uv-plane bounding a domain A_1. It is essential that there should be a 1–1 correspondence between points of A in the xy-plane and points of A_1 in the uv-plane. Further let t be a parameter that steadily increases or decreases from t_0 to t_1 as (x, y) describes the curve C positively; then, by reason of the 1–1 mapping, (u, v) will describe the curve C_1 as t steadily increases or decreases from t_0 to t_1, but it will not necessarily describe the curve C_1 positively.

We shall assume that x, y, u, v are single-valued differentiable functions of t.

Then A is given by the line integral

$$A = \int_C x \, dy = \int_{t_0}^{t_1} x \frac{dy}{dt} \, dt.$$

Now, since y is a function of u and v, and u and v are each functions of t,

$$\frac{dy}{dt} = \frac{\partial y}{\partial u}\frac{du}{dt} + \frac{\partial y}{\partial v}\frac{dv}{dt}.$$

Hence

$$A = \int_{t_0}^{t_1} x \left(\frac{\partial y}{\partial u}\frac{du}{dt} + \frac{\partial y}{\partial v}\frac{dv}{dt} \right) dt$$

$$= \pm \int_{C_1} x \frac{\partial y}{\partial u} \, du + x \frac{\partial y}{\partial v} \, dv,$$

the ambiguous sign being positive if C_1 is described positively as t changes from t_0 to t_1, and negative if C_1 is described negatively.

On transforming the line integral into a double integral using Green's theorem,

$$A = \pm \iint_{A_1} \left\{ \frac{\partial}{\partial u}\left(x\frac{\partial y}{\partial v}\right) - \frac{\partial}{\partial v}\left(x\frac{\partial y}{\partial u}\right) \right\} du \, dv$$

$$= \pm \iint_{A_1} \left\{ \left(\frac{\partial x}{\partial u}\frac{\partial y}{\partial v} + x\frac{\partial^2 y}{\partial u \, \partial v} \right) - \left(\frac{\partial x}{\partial v}\frac{\partial y}{\partial u} + x\frac{\partial^2 y}{\partial u \, \partial v} \right) \right\} du \, dv$$

$$= \pm \iint_{A_1} \left(\frac{\partial x}{\partial u}\frac{\partial y}{\partial v} - \frac{\partial x}{\partial v}\frac{\partial y}{\partial u} \right) du \, dv.$$

DOUBLE INTEGRALS

The expression $\dfrac{\partial x}{\partial u}\dfrac{\partial y}{\partial v} - \dfrac{\partial x}{\partial v}\dfrac{\partial y}{\partial u}$ or

$$\begin{vmatrix} \dfrac{\partial x}{\partial u} & \dfrac{\partial y}{\partial u} \\ \dfrac{\partial x}{\partial v} & \dfrac{\partial y}{\partial v} \end{vmatrix}$$

is called the Jacobian of x, y with respect to u, v and is denoted by $\dfrac{\partial(x, y)}{\partial(u, v)}$; we shall denote it briefly by J. Thus

$$\iint_A dx\, dy = \pm \iint_{A_1} J\, du\, dv,$$

the ambiguous sign being positive if C and C_1 are described in the same direction and negative if C and C_1 are described in different directions. It will be found that J is positive if C and C_1 are described in the same directions and negative otherwise. So we may write

$$\iint_A dx\, dy = \iint_{A_1} |J|\, du\, dv.$$

Example. For the first problem on mapping given on p. 206

$$J = \begin{vmatrix} \dfrac{\partial x}{\partial u} & \dfrac{\partial y}{\partial u} \\ \dfrac{\partial x}{\partial v} & \dfrac{\partial y}{\partial v} \end{vmatrix} = \begin{vmatrix} 1 & 1 \\ 1 & -1 \end{vmatrix} = -2,$$

which agrees with the statement made above that if C and C_1 are described in opposite directions J is negative. Hence

$$\iint_A dx\, dy = \iint_{A_1} 2\, du\, dv,$$

i.e. $A = 2A_1$ which is otherwise clear.

Example. In the transformation from Cartesian to polar coordinates

$$x = r\cos\theta, \qquad y = r\sin\theta,$$

$$\dfrac{\partial(x, y)}{\partial(r, \theta)} = \begin{vmatrix} \cos\theta & \sin\theta \\ -r\sin\theta & r\cos\theta \end{vmatrix} = r.$$

Hence

$$\iint_A dx\, dy = \iint_{A_1} r\, dr\, d\theta.$$

Exercises 7(g)

1. A triangle with vertices $(0, 0)$, $(1, 2)$, $(2, 1)$ in the xy-plane is mapped into the uv-plane by the transformation
$$x = 3u - v,$$
$$y = u + 2v.$$
Find the domain in the uv-plane and consider the directions in which the boundaries of the xy-domain and the uv-domain are described. Find J and hence establish a relation between the area of the triangle in the xy-plane and the mapped domain in the uv-plane. Also prove this relation by elementary methods.

2. Prove that the area of the simple closed curve $f\left(\dfrac{x}{a}, \dfrac{y}{b}\right) = 0$ is ab times the area of the curve $f(x, y) = 0$.

3. If $x = a \sin \theta \sin \phi$, $y = b \cos \theta \sin \phi$, are the mapping transformations, prove that
$$\iint dx\, dy = \iint ab \sin \phi \cos \phi\, d\theta\, d\phi.$$

4. If $f(x, y, u, v) = 0$ and $\phi(x, y, u, v) = 0$, find the Jacobian of u, v with respect to x, y.

5. If $u \cos x + v \sin x = 1$, and $u \cos y + v \sin y = 1$, prove that
$$\frac{\partial(x, y)}{\partial(u, v)} = \pm \frac{\sin(x - y)}{u^2 + v^2 - 1}.$$

6. A parallelogram in the xy-plane with vertices $(1, 3)$, $(1, -1)$, $(2, 2)$, and $(0, 0)$ is mapped onto the uv-plane using the transformation
$$u = x + y + 2,$$
$$v = 3x - y + 2.$$
Establish a relation between the area of the parallelogram and the area mapped into the uv-plane. Also establish this relation by elementary methods.

7. z and w are functions of u and v, and u and v are functions of x and y. Prove that
$$\frac{\partial(z, w)}{\partial(x, y)} = \frac{\partial(z, w)}{\partial(u, v)} \frac{\partial(u, v)}{\partial(x, y)}.$$

We shall now establish a formula for the change of variables in a double integral. Let
$$I = \iint_A F(x, y)\, dx\, dy$$
and let
$$x = f(u, v),$$
$$y = \phi(u, v),$$

be a 1–1 mapping of the domain A in the xy-plane into the domain A_1 in the uv-plane.

$$\iint_A F(x, y)\,dx\,dy = \lim_{\substack{\delta x \to 0 \\ \delta y \to 0}} \sum \sum F(x, y)\,\delta x\,\delta y.$$

Now from the result proved above that $\iint_A dx\,dy = \iint_{A_1} |J|\,du\,dv$, it follows that
$$\delta x \delta y = |J|\,\delta u \delta v,$$
and hence

$$\iint_A F(x, y)\,dx\,dy = \lim_{\substack{\delta x \to 0 \\ \delta y \to 0}} \sum \sum F(x, y)\,|J|\,\delta u\,\delta v$$

$$= \lim_{\substack{\delta u \to 0 \\ \delta v \to 0}} \sum \sum F(f(u, v), \phi(u, v))\,|J|\,\delta u\,\delta v$$

$$= \iint_{A_1} F(f(u, v), \phi(u, v))\,|J|\,du\,dv.$$

Thus
$$\iint_A F(x, y)\,dx\,dy = \iint_{A_1} F(f(u, v), \phi(u, v))\,|J|\,du\,dv.$$

The conditions for the validity of this important result are that there must be a 1–1 correspondence between the domains A and A_1 and that the Jacobian J must be of fixed sign throughout the domains.

Example. A solid has for base the parallelogram A with vertices $P(1, 1)$, $Q(0, 2)$, $R(-1, -1)$ and $S(0, -2)$. The sides of the solid are perpendicular to the base and the face opposite the base is a portion of the surface $z = x^2(x+y)^2$. The volume of the solid is

$$\iint_A x^2(x+y)^2\,dx\,dy.$$

This integral can be calculated straightforwardly or by making a change of variables. If we apply the transformations

$$u = x+y,$$
$$v = 3x-y,$$

the area $PQRS$ is mapped into A_1 with vertices $P_1(2, 2)$, $Q_1(2, -2)$, $R_1(-2, -2)$, and $S_1(-2, 2)$.

$$x = \tfrac{1}{4}(u+v), \qquad y = \tfrac{1}{4}(3u-v),$$

whence
$$J = \begin{vmatrix} \tfrac{1}{4} & \tfrac{3}{4} \\ \tfrac{1}{4} & -\tfrac{1}{4} \end{vmatrix} = -\tfrac{1}{4}$$

Hence J has a fixed sign which is negative, corresponding to the fact that $PQRS$ is anticlockwise, while $P_1Q_1R_1S_1$ is clockwise. It is further clear that there is a 1–1 correspondence between the domains A and A_1. Hence the volume of the solid is

$$\iint_{A_1} \tfrac{1}{16}(u+v)^2 u^2 (\tfrac{1}{4}) \, du \, dv = \tfrac{1}{64} \int_{-2}^{2} u^2 \, du \int_{-2}^{2} (u+v)^2 \, dv = \tfrac{56}{45}.$$

Exercises 7(h)

1. If $x = r \cos \theta$, $y = r \sin \theta$, prove that if A_1 on the $r\theta$ plane is a 1–1 mapping of A on the xy-plane, then

$$\iint_A f(x, y) \, dx \, dy = \iint_{A_1} f(r \cos \theta, r \sin \theta) r \, dr \, d\theta.$$

A is the region in the xy-plane bounded by the circles

$$x^2 + y^2 = 1, \qquad x^2 + y^2 = 4.$$

Evaluate $\iint_A (x^2+y^2)^{\frac{1}{2}} \, dx \, dy$. Also evaluate $\iint_A (x^2+y^2)^{\frac{3}{2}} \, dx \, dy$.

2. A is the region in the xy-plane defined by $x \geq 0$, $y \geq 0$, $x+y \leq 1$. Prove that

$$\iint_A xyf(x+y) \, dx \, dy = \frac{1}{6} \int_0^1 u^3 f(u) \, du.$$

3. A uniform solid of uniform density ρ occupies the region lying inside the circular cylinder $x^2 + y^2 = a^2$ and between the plane $z = 0$ and the surface $z = x^2 y^2$. Express its moment of inertia about OZ as a double integral of the form

$$\iint f(x, y) \, dx \, dy.$$

Evaluate this double integral directly and also by transforming the integral into polar coordinates.

4. A is the quadrilateral in the xy-plane, vertices $(0, 1)$, $(1, 4)$, $(3, 5)$, and $(2, 2)$. Evaluate the integral

$$\iint_A (2y-x-2)(3x-y+1) \, dx \, dy$$

DOUBLE INTEGRALS

in two ways (i) as a double integral in x and y, (ii) by changing the variables to u, v where
$$2y - x - 2 = 5u,$$
$$3x - y + 1 = 5v.$$

5. R is the region $a^2 \leq x^2 + y^2 \leq b^2$, $b > a > 0$. Prove that
$$\iint_R x^2 \, dx \, dy = \frac{\pi}{4}(b^4 - a^4).$$

We shall conclude this chapter by investigating an integral of some importance in statistics and science,
$$\int_0^\infty e^{-x^2} \, dx.$$

Let
$$I = \int_0^k e^{-x^2} \, dx,$$

where $k > 0$. Then
$$I^2 = \left(\int_0^k e^{-x^2} \, dx\right)\left(\int_0^k e^{-y^2} \, dy\right),$$

and the repeated integral can be expressed as a double integral
$$\iint_A e^{-(x^2+y^2)} \, dx \, dy,$$

where A is the square defined by $0 \leq x \leq k$, $0 \leq y \leq k$. Now change the variables into polar coordinates
$$x = r \cos \theta, \qquad y = r \sin \theta,$$

where $r \geq 0$, $0 \leq \theta \leq \frac{1}{2}\pi$. Then
$$I^2 = \iint_{A_1} e^{-r^2} r \, dr \, d\theta,$$

where A_1 is the area in the $r\theta$-plane corresponding to A in the xy-plane. It is clear that the portion of A for which $r \leq k$ will map into a rectangle A_2 in the $r\theta$ plane defined by
$$0 \leq r \leq k, \qquad 0 \leq \theta \leq \tfrac{1}{2}\pi.$$

Now the maximum value of r for points of A is $k\sqrt{2}$, and the portion of A for which $r > k$ will map into a portion of the rectangle

in the $r\theta$-plane defined by
$$k \le r \le k\sqrt{2}, \quad 0 \le \theta \le \tfrac{1}{2}\pi.$$
Hence, since the integrand is positive, $\iint_{A_1} e^{-r^2} r \, dr \, d\theta$ will lie between
$$\int_0^k e^{-r^2} r \, dr \int_0^{\frac{\pi}{2}} d\theta \quad \text{and} \quad \int_0^{k\sqrt{2}} e^{-r^2} r \, dr \int_0^{\frac{\pi}{2}} d\theta, \quad \text{i.e. between} \quad \frac{\pi}{4}(1-e^{-k^2}) \quad \text{and}$$
$\frac{\pi}{4}(1-e^{-2k^2})$. Now, as k tends to infinity, both these limits tend to $\pi/4$.
Hence as k tends to infinity, $I^2 \to \frac{\pi}{4}$ and $I \to \frac{\sqrt{\pi}}{2}$. It follows that
$$\int_0^\infty e^{-x^2} dx = \frac{\sqrt{\pi}}{2}.$$

Miscellaneous problems on Chapter 7

1. A is a domain in the xy-plane bounded by the quadrilateral vertices $(4, 4)$, $(2, 0)$, $(-4, 0)$, and $(-1, 3)$. Evaluate the double integral
$$\iint_A (x+y-2)^2(x-2y+4) \, dx \, dy,$$
by mapping the region A into the uv-plane using the transformations $x+y-2 = 6u$, $x-2y+4 = 3v$.

2. C is the closed curve bounded by the parabolas $y = x^2$, $x = y^2$, described positively. Find the value of the line integral
$$\int_C (x^2+2xy) \, dx + (y^2-x) \, dy.$$
Verify Green's theorem in this case.

3. A is the domain defined by
$$-2a \le x+y \le 2a,$$
$$-2a \le x-y \le 2a;$$
prove that
$$\iint_A \phi(x+y)\phi(x-y) \, dx \, dy = 2\left[\int_{-a}^a \phi(2x) \, dx\right]^2.$$

4. A is the triangular domain bounded by the lines $x = 0$, $y = 0$, and $x+y = k$. Prove that the value of the double integral
$$\iint_A x^p y^q \, dx \, dy,$$

where p and q are positive integers, is
$$\frac{p!q!}{(p+q+2)!}k^{p+q+2}.$$

5. A denotes the interior of the ellipse with equation
$$x^2+4y^2-4x+12y+12 = 0.$$
Evaluate the double integral $\iint_A (y^2+y+2)\,dx\,dy$.

6. Evaluate the double integral $\iint_R x^7 y^3 e^{x^2 y^2}\,dx\,dy$, where R is the region defined by $0 \le y \le 2x \le 2$, $xy \ge 1$.

7. D is the domain defined by $x+y \le 1$, $x \ge 0$, $y \ge 0$. Prove that, with suitable restrictions,
$$\iint_D x^p y^q f(x+y)\,dx\,dy = \frac{p!q!}{(p+q+1)!}\int_0^1 u^{p+q+1}f(u)\,du.$$

8. Prove that the area of the portion of the surface of the sphere
$$x^2+y^2+z^2 = 1$$
which lies inside the cylinder $x^2+y^2-y = 0$ is $2\pi-4$.

9. Prove that the area of a loop of the curve $x = a\sin t$, $y = a\sin 2t$ is $\tfrac{4}{3}a^2$.

10. Evaluate
$$\iint_R (x+3y)^3\,dx\,dy,$$
   (i) when R is the triangular region bounded by the lines $x = 0$, $y = 0$, $x+3y = 1$,
   (ii) when R is the region defined by $x^2+y^2 \le 4$, $x \le 0$.

11. Sketch the curve $x^5+y^5 = 5ax^2y^2$ and find the area of the loop. Repeat the problem for the curve $x^7+y^7 = 7ax^3y^3$.

12. Prove that the volume of the portion of the circular cylinder $x^2+y^2 = a^2$ lying between the plane $z = 0$ and the surface $z = (x+y+a)^2$ is $\tfrac{3}{2}\pi a^4$. Prove also that the z-coordinate of the mean centre of the solid is $\tfrac{3}{2}a^2$.

13. If x, y, z are functions of u and v with continuous first order partial differential coefficients, show that, if z is regarded as a function of x and y,
$$\frac{\partial z}{\partial x} = -\frac{\partial(y,z)}{\partial(u,v)}\bigg/\frac{\partial(x,y)}{\partial(u,v)},$$
and find a similar expression for $\dfrac{\partial z}{\partial y}$.

Find $\dfrac{\partial z}{\partial x}$ in terms of u and v if $x = u^2+v$, $y = u+v^2$, $z = e^{u^2+v^2}$.

14. $PQRS$ is a parallelogram in the xy-plane with vertices $P(1, 2)$, $Q(2, 3)$, $R(1, 0)$, $S(0, -1)$. A solid has base $PQRS$, sides perpendicular to the base, and is bounded by the surface $z = x^2+3y^2$. Express the volume of the solid as an integral in x and y and hence find the volume.

Find a pair of linear transformations that map the parallelogram into a square in the uv-plane with vertices $(0, 0)$, $(1, 1)$, $(2, 0)$, $(1, -1)$. Express the volume of the solid as a double integral in u and v and hence find the volume.

15. If u and v are functions of x and y, then, in general, there is not a relation of the form $f(u, v) = 0$, independent of x and y. Prove that a necessary condition for such a relation between u and v is

$$\frac{\partial(u, v)}{\partial(x, y)} = 0.$$

It may be proved that the condition is also sufficient.

16. Sketch the curve $y^2(a+x) = x^2(3a-x)$. Show that the coordinates of any point on the curve may be expressed in the form $x = \dfrac{a \sin 3t}{\sin t}$, $y = \dfrac{a \sin 3t}{\cos t}$ and prove that the area of the loop of the curve is $3\sqrt{3}a^2$.

17. A hyperbolic paraboloid and a circular cylinder are given by the equations $z = x^2-y^2$, $x^2+y^2-6x+5 = 0$. Find the volume inside the cylinder between the paraboloid and the plane $z = 0$.

18. If u and v are functions of x and y, and $J = \dfrac{\partial(u, v)}{\partial(x, y)}$, prove that

$$\frac{\partial(J, v)}{\partial(x, y)} = J\frac{\partial J}{\partial u}.$$

19. Prove that the area of that portion of the surface $2z = x^2+y^2$ which lies within the cylinder

$$x^2+y^2-8x+12 = 0$$

can be represented in the form

$$\iint_A \sqrt{(1+x^2+y^2)}\, dx\, dy,$$

where A is the domain $x^2+y^2-8x+12 \leq 0$, $z = 0$. Change the variables in the double integral from x, y to r, θ, where $x = r \cos \theta$, $y = r \sin \theta$.

DOUBLE INTEGRALS

Hence show that the required area is

$$\frac{2}{3}\int_0^{\frac{\pi}{6}} \{(1+r_2^2)^{\frac{3}{2}} - (1+r_1^2)^{\frac{3}{2}}\}\, d\theta,$$

where r_1 and r_2 are the roots of the equation $r^2 - 8r\cos\theta + 12 = 0$.

20. A is the domain bounded by the parallelogram with vertices $(0, 0)$, $(1, 1)$, $(2, 3)$, $(1, 2)$ in the xy-plane. Evaluate

$$\iint_A (y-x)^2(y-2x)\, dx\, dy,$$

(i) as a double integral in x and y; (ii) by changing the variables to u, v where $y-x = u$, $y-2x = v$. Show that the value of the integral is $-\frac{1}{6}$.

21. Evaluate the integral $\iint_R (x^2+y^2)\, dx\, dy$, where R is the region

$$\frac{x^2}{a^2} + \frac{y^2}{b^2} \le 1.$$

22. $I = \iint_R \dfrac{ky \cos kr_1 \cos kr_2}{r_1 r_2}\, dx\, dy$, where k is a constant,

$$r_1 = \{(x+1)^2 + y^2\}^{\frac{1}{2}}, \qquad r_2 = \{(x-1)^2 + y^2\}^{\frac{1}{2}},$$

and R is the upper half of the ellipse foci $(1, 0)$, $(-1, 0)$ and major axis, length 4.

Change the variables of integration from (x, y) to (u, v), where

$$u = \tfrac{1}{2}(r_1+r_2), \qquad v = \tfrac{1}{2}(r_1-r_2),$$

and hence show that
$$I = 4 \sin k \cos^3 k.$$

23. Calculate the volume and the surface area of that part of the sphere of radius $2a$, centre the origin, which is cut off by the cylinder

$$x^2 + y^2 - 2ax = 0.$$

24. If x and y are functions of u and v giving a 1–1 relation between a domain in the xy-plane and a corresponding domain in the uv-plane, write down the relation between the area of a small region $\delta x \delta y$ in the xy-plane and the corresponding small area $\delta u \delta v$ in the uv-plane.

Express independently a similar result for $\delta u \delta v$ in terms of $\delta x \delta y$. Hence, prove that

$$\frac{\partial(x, y)}{\partial(u, v)} \frac{\partial(u, v)}{\partial(x, y)} = 1.$$

Also prove the result directly from the analytical definition of a Jacobian.

25. If U, V are functions of u, v and u, v are functions of x, y prove (i) by considering corresponding small areas in the UV, uv, xy planes, (ii) by using the analytical definitions of the terms that

$$\frac{\partial(U, V)}{\partial(x, y)} = \frac{\partial(U, V)}{\partial(u, v)} \frac{\partial(u, v)}{\partial(x, y)}.$$

26. If $y_1, y_2, y_3, z_1, z_2, z_3$ are functions of x_1, x_2, x_3 find an expression for $\dfrac{\partial(z_1, z_2, z_3)}{\partial(y_1, y_2, y_3)}$ in terms of $\dfrac{\partial(y_1, y_2, y_3)}{\partial(x_1, x_2, x_3)}$ and $\dfrac{\partial(z_1, z_2, z_3)}{\partial(x_1, x_2, x_3)}$.

If $y_1 = x_1^2 + x_2^2 + x_3^2$, $y_2 = x_1^3 + x_2^3 + x_3^3$, $y_3 = x_1^4 + x_2^4 + x_3^4$, $z_1 = x_1 + x_2 + x_3$, $z_2 = x_1 x_2 + x_2 x_3 + x_3 x_1$, $z_3 = x_1 x_2 x_3$, prove that the value of $\dfrac{\partial(z_1, z_2, z_3)}{\partial(y_1, y_2, y_3)}$ when $x_1 = 1$, $x_2 = 2$, $x_3 = 3$ is $-\frac{1}{72}$.

27. A space curve C is defined by $u(x, y, z) = 0$, $v(x, y, z) = 0$. If s is the length of the arc of the curve and $f(s)$ is any function of s, prove that

$$\frac{df}{ds} = \frac{1}{\lambda} \sum \frac{\partial f}{\partial x} \frac{\partial(u, v)}{\partial(y, z)},$$

where

$$\lambda^2 = \left\{\frac{\partial(u, v)}{\partial(y, z)}\right\}^2 + \left\{\frac{\partial(u, v)}{\partial(z, x)}\right\}^2 + \left\{\frac{\partial(u, v)}{\partial(x, y)}\right\}^2.$$

28. Prove that the volume cut off in the octant defined by $x, y, z \geq 0$ from the cylinder $x^2 + (y-a)^2 = b^2$ by the surface $xy = z$ is $\frac{2}{3}ab^3$.

29. A volume is bounded by the plane $z = 0$, the cylinder $x^2 + y^2 = 2ax$, $z \geq 0$, and the cone $4z^2 = x^2 + y^2$. Calculate the volume and the area of that part of the surface bounded by the cone.

30. Prove that the area of that part of the surface $z^2 = 2xy$ which lies inside the sphere $x^2 + y^2 + z^2 = 1$ is $\pi/\sqrt{2}$.

31. Prove that the area in the positive quadrant of the xy-plane bounded by the curves

$$y^2 = x, \quad y^2 = 4x, \quad xy^2 = 4, \quad xy^2 = 9,$$

is $(2\sqrt{6} - \frac{8}{3})(\sqrt{2}-1)$ by two methods:
(i) by expressing the area as a double integral in x and y,
(ii) by changing the variables in this double integral using the transformations $\dfrac{y^2}{x} = u$, $xy^2 = v$.

32. If R is the elliptic region defined by

$$\frac{x^2}{a^2} + \frac{y^2}{b^2} \leq 1,$$

where a and b are positive, prove that $\iint_R (x^2+y^2)\,dx\,dy = \frac{1}{4}\pi ab(a^2+b^2)$.

Obtain the result by two methods (i) as a double integral in x and y, (ii) by changing the variables to r and θ where

$$x = ar\cos\theta, \qquad y = br\sin\theta.$$

33. Prove that the volume bounded by the plane $z = 0$, the upper half of the cone $x^2+y^2 = z^2$ and the cylinder $x^2+y^2 = 4y$ is $\frac{256}{9}$.

34. Prove that, if C is the region defined by

$$x^2+y^2-4y \leq 0, \qquad x \geq 0,$$

$$\iint_C \frac{x^3 y^3}{(x^2+y^2)^2} e^{(x^2+y^2)}\,dx\,dy = \frac{23}{2048}e^{16} + \frac{235}{6144}.$$

8

The Symbolism of Modern Algebra

The preceding chapters of this volume have been written in the traditional language of mathematics, but over the last few years there has occurred something of a revolution in mathematical thought with a growing emphasis on mathematical structure rather than mathematical techniques and the introduction of a new symbolism which has applications outside the sphere of traditional mathematics.

It will be the purpose of this chapter to give a brief account of that symbolism with an elementary treatment of sets.

Sets

It is assumed that the reader is familiar with the concept of a set and the use of the Venn diagram in representing sets.

We can define a particular set by listing all the members of the set, e.g.
$$A = \{1, 2, 5, 9\},$$
or by stating properties which define the elements of the set, e.g.
$$B = \{x \mid x \text{ is odd}\},$$
which defines the infinite set $\{1, 3, 5, 7 \cdots\}$.

If x is a member of a set A we write $x \in A$; if x is not a member of the set A we write $x \notin A$. The order of elements in a set is not relevant, e.g.
$$\{1, 2, 3\} = \{2, 1, 3\};$$
if on occasions, for example, in defining the coordinates of a point, we wish the order of terms to be relevant we would refer to the set as an ordered set and write it as (1, 2, 3), using parentheses brackets instead of braced brackets, which are always used for non-ordered sets. Note that $(1, 2, 3) \neq (2, 1, 3)$.

It is possible to state a property defining a set which is not satisfied by any element, e.g.

$$\{x \mid x \text{ is his own father}\};$$

we refer to such a set as an empty set { } and denote it by ϕ.

Subsets

If N is a set consisting of some of the elements of a set A, we call N a subset of A, and write $N \subset A$; it is convenient to include A itself among the subsets of A, but some authors refer to A as an improper subset and other subsets as proper subsets. For example (i) $\{a\}$ and $\{a, c\}$ are subsets of $\{a, b, c\}$. (ii) If L is the set of points on a straight line lying in a plane α, and C is the set of points on a circle in α, then L and C are subsets of the set of points forming the plane α.

Clearly two sets A and B are equal if $A \subset B$ and $B \subset A$.

Exercises 8(a)

1. Prove that if $A \subset B$ and $B \subset C$, then $A \subset C$.
2. Under what conditions are the following statements true?

    (i) $\{a, b\} = \{c, d\}$ (ii) $(a, b) = (c, d)$.

3. If A is a set of n elements, how many subsets of A are there, including A itself?
4. Give some examples of properties defining membership of a set which yield empty sets.
5. $A = \{x \mid x \text{ is a square}\}$; $B = \{x \mid x \text{ is a rectangle}\}$;

    $C = \{x \mid x \text{ is a quadrilateral}\}$; $D = \{x \mid x \text{ is a trapezium}\}$.

Find which of the sets, if any, are subsets of (i) A, (ii) B, (iii) C, (iv) D.

Operations on sets

The *intersection* of two sets A and B is defined as the set of elements common to A and B and is denoted by $A \cap B$. For example,

$$\{a, b, c\} \cap \{b, d, e\} = \{b\}.$$

If $A \cap B = \phi$, we call the sets A and B *disjoint*.

The *union* of two sets A and B is defined as the set of elements which are members *either* of A or B or of both, and is denoted by $A \cup B$. It follows that if $x \in A$, then $x \in A \cup B$, and if $y \in B$, then $y \in A \cup B$. For example,

$$\{a, b, c\} \cup \{b, d, e\} = \{a, b, c, d, e\}.$$

Before considering the laws governing intersection and union of sets, it is convenient to introduce a symbolism for 'implies'.

If ABC is a triangle, the condition $AB = AC$ implies that
$$\angle B = \angle C.$$
We shall write this fact in the form
$$AB = AC \Rightarrow \angle B = \angle C,$$
the symbol $\Rightarrow$ meaning that the condition on the arrow-head side of the symbol is a logical deduction from the condition preceding the symbol. We also introduce the symbol $\Leftarrow$ meaning that the condition on the arrow-head side of the symbol is a logical deduction from the condition following the symbol.

If r and s are two conditions, the fact that $r \Rightarrow s$ does not imply that $r \Leftarrow s$.

If ABC is a triangle, the condition $\angle B = \angle C$ implies that $AB = AC$, i.e.
$$\angle B = \angle C \Rightarrow AB = AC.$$

Combining this result with that above we have
$$AB = AC \Rightarrow \angle B = \angle C,$$
$$AB = AC \Leftarrow \angle B = \angle C.$$
These two statements are combined into the single statement
$$AB = AC \Leftrightarrow \angle B = \angle C.$$

In general, if r and s are two conditions, the statement $r \Leftrightarrow s$ means that r implies s and s implies r. We then say that the conditions r and s are equivalent.

Further symbols used are : as an alternative to the symbol | to denote 'such that' and $\exists$ to denote 'there exist(s)'. For example, if A is the set of integers $A = \{1, 2, 3,...\}$ and $x:2x+5 = 19$, then $\exists x \in A$. The symbol $\forall$ denotes 'for all relevant values of'.

The universal set

In any particular problem involving a number of sets it is convenient to have a notation for a single set of which the given sets are subsets. Such a set is called the *universal set* and is denoted by $\mathscr{E}$. For example,

(i) in a geometrical problem in which we are concerned with sets of points on given surfaces we could take as $\mathscr{E}$ the set of points in space;

(ii) in a problem on distribution of animals throughout the world we could take as $\mathscr{E}$ the set of animals in the world.

If A is a subset of the universal set $\mathscr{E}$, the complement of the set A, denoted by A', is the set of elements of $\mathscr{E}$ which are not members of the set A. For example, if $\mathscr{E}$ is the set of integers, and A is the set of prime numbers, then A' is the set of non-prime numbers.

One important universal set is the set of real numbers, denoted† by $R^{\#}$.

Definition of a function

(i) Consider the set $A = \{x \mid x \in R^{\#}\}$. If we construct a second set $B = \{y : y = x^2\}$, then to each element of A there corresponds a unique element of B.

(ii) Consider the set $A = \{x \mid x$ is a natural number$\}$. If we construct a second set $B = \{y : y = x$th prime number$\}$, then to each element of A there corresponds a unique element of B.

(iii) Consider the set $A = \{x \mid x$ is a child$\}$. If we construct a second set $B = \{y \mid y$ is the father of $x\}$, then to each element of A there corresponds a unique element of B.

If for each element of a set A we can assign a unique element of a set B, we call the assignment a function and write $f : A = B$. If x is any element of set A, and y the corresponding element of set B, we call y the image of x and we write $y = f(x)$.

The set A is called the domain of the function f and the set B the codomain. We say that A is mapped onto B by the function f and write

$$A \xrightarrow{f} B \quad \text{or} \quad f : A \longrightarrow B.$$

In modern mathematics, the term function is reserved for mappings from set A to set B for which each element of A has a unique image in B, which implies that the relation between elements of set A and set B is either many–one or one–one, but not one–many. This is a difference in notation from traditional mathematics, where the term

---

† There is no universal agreement about the notation to be used for frequently occurring sets, such as the set of real numbers. Some authors use R for the set of real numbers, R^+ for the set of positive real numbers and R^- for the set of negative real numbers.

function is used also for a one-many relationship, the function then being called a many-valued function.

If the relation between A and B is a one-one relation so that to each element of A there corresponds a unique element of B, and to each element of B there corresponds a unique element of A, we have an assignment such that for each element of set B we can assign a unique element of set A. This assignment is a function which we denote by f^{-1}, called the inverse of the function f. We have therefore the relations

$$f:A = B,$$
$$f^{-1}:B = A,$$

or in mapping terms

$$A \xrightarrow{f} B,$$
$$B \xrightarrow{f^{-1}} A.$$

A function f which possesses an inverse is called a bijective function.

Exercises 8(b)

1. State for which of the following functions there is an inverse function:
    (i) $A \xrightarrow{f} B$, where A is R and $f(x) = x^3$;
    (ii) $A \xrightarrow{f} B$, where A is R and $f(x) = x^2$;
    (iii) $A \xrightarrow{f} B$, where A is the set of complex numbers and $f(x) = x^3$.

2. (i) A and B are two sets with at least one common element. Represent A, B, $A \cup B$, $A \cap B$ by shaded Venn diagrams.
    (ii) A and B are two disjoint sets. Represent A, B, $A \cup B$ by shaded Venn diagrams. What is the value of $A \cap B$ in this case?

3. If the universal set is the set of all natural numbers and A is the set of all odd numbers, what is A'?
    If $A = \{x \mid x \in R^\#, x \text{ is rational}\}$, what is A'?

4. Prove that
    (i) $(A')' = A$,
    (ii) $A \cup A' = \mathscr{E}$,
    (iii) $A \cap A' = \phi$,
    (iv) $x \in A \Leftrightarrow x \notin A'$.

5. If Z is the set of all natural numbers and
$$A = \{x \mid x = 3t+1 \text{ or } 3t+2, t \in Z\},$$
find the property defining A'.

6. If A and B are two possible outcomes of an experiment and $p(A)$ and $p(B)$ are the probabilities of the two events, prove that

$$p(A \cup B) = p(A) + p(B) - p(A \cap B).$$

The algebra of sets

The following laws are fundamental in the algebra of sets.
   (i) *The idempotent laws.* $A \cup A = A$; $A \cap A = A$.
   (ii) *The commutative laws.* $A \cup B = B \cup A$; $A \cap B = B \cap A$.
   (iii) *The associative laws.*

$$(A \cup B) \cup C = A \cup (B \cup C); \quad (A \cap B) \cap C = A \cap (B \cap C).$$

(iv) *The distributive laws.*

$$A \cup (B \cap C) = (A \cup B) \cap (A \cup C);$$
$$A \cap (B \cup C) = (A \cap B) \cup \cup (A \cap C).$$

We shall prove one of these theorems and leave the proof of the other theorems as an exercise for the reader, who may find Venn diagrams useful in organizing the proof.

THEOREM. $(A \cup B) \cup C = A \cup (B \cup C)$.
PROOF. $(A \cup B) \cup C = \{x \mid x \in A \cup B \text{ or } x \in C\}$

$$\Rightarrow \{x \mid \overline{x \in A \text{ or } x \in B} \text{ or } x \in C\}$$
$$\Rightarrow \{x \mid x \in A \text{ or } \overline{x \in B \text{ or } x \in C}\}$$
$$\Rightarrow \{x \mid x \in A \text{ or } x \in B \cup C\}$$
$$\Rightarrow \{x \mid x \in A \cup (B \cup C)\}.$$

It follows that $(A \cup B) \cup C \subset A \cup (B \cup C)$. We prove in a similar fashion that $A \cup (B \cup C) \subset (A \cup B) \cup C$. These two results imply that

$$(A \cup B) \cup C = A \cup (B \cup C).$$

Other standard results which the reader should illustrate by the use of Venn diagrams are

$$A \cup \phi = A; \quad A \cup \mathscr{E} = \mathscr{E}; \quad A \cap \phi = \phi; \quad A \cap \mathscr{E} = A;$$
$$(A')' = A; \quad (A \cup B)' = A' \cap B'; \quad (A \cap B)' = A' \cup B'.$$

Exercises 8(c)

1. Illustrate the following theorems by the use of Venn diagrams and prove them formally.
$$A \cup (A \cap B) = A; \quad A \cap (A \cup B) = A.$$

The structure of an operation

If a and b are elements of $R^\#$, we can combine a and b in various ways, e.g. $a+b$, $a-b$, ab. In each case the result of the operation is an *element* of $R^\#$.

If a and b are elements of the set of natural numbers, the operations $a+b$, ab a^b yield members of the set but the operation $a-b$ does not do so unless $a > b$.

If a and b are elements of the set of odd integers, the operation $a+b$ does not yield a member of the set, but the operation ab does.

If **P** and **Q** are members of the set of three-dimensional vectors, the operation **P**+**Q**, and the vector product $P \wedge Q$ yield members of the set, but the scalar product **P.Q** does not do so.

If A and B are two sets, the operations $A \cup B$ and $A \cap B$ yield other sets.

In general, if a and b are elements of the set S, and there is a law of combination, denoted by $*$, giving a unique $a*b$, we call the operation a binary operation. An important province of modern mathematics is the investigation of the structure of the operation in relation to the elements of S. By structure we understand the laws governing the operation. Note that there are two factors in the nature of structure, the operation and the set of elements to which the operation is applied.

If for any pair of elements a and b of the set S, $a*b$ is also a member of S, we say that the set S is closed in respect of the operation $*$.

For example, the set $R^\#$ is closed in respect of addition; the set of natural numbers is closed in respect of multiplication; the set of prime numbers is not closed in respect of addition and not closed in respect of multiplication.

Expressing the condition for closure symbolically, we say that the set is closed in respect of the operation $*$ if
$$\forall a, b \in S \quad a*b \in S.$$

The operation $*$ is called associative if $(a*b)*c = a*(b*c)$, and commutative if $a*b = b*a$.

Neutral† elements of a set

In the set of real numbers, two elements, 0 and 1, occupy a special position. If a is any member of the set, then

$$a+0 = 0+a = a.$$

Thus, in respect of the operation of addition, 0 has a neutral role in the sense of the results stated. Again, if a is any member of the set,

$$a.1 = 1.a = a.$$

Thus, in respect of the operation of multiplication, 1 has a neutral role in the sense of the results stated.

In general, if for a set S there exists an element e such that for the operation $*$

$$a * e = e * a = a,$$

for all elements a of S, we call e a neutral element (or identity element) of the set in respect of the operation $*$. It will be clear that in respect of any particular operation $*$, there can be only one neutral element. For if e_1 and e_2 were neutral elements,

$$e_1 * e_2 = e_2,$$

and

$$e_2 * e_1 = e_1;$$

but $e_1 * e_2 = e_2 * e_1$ and hence $e_1 = e_2$.

Not all sets have a neutral element for a particular operation. The set of even natural numbers, for example, has no neutral element in respect of the operation of addition, and no neutral element in respect of the operation of multiplication.

Inverse operations

For the set $R^\#$, there is associated with every element x of the set, other than 0, an inverse element y such that $xy = 1$. y is called the inverse of x in respect of the operation of multiplication. We write $y = 1/x$ or x^{-1} and call the operation of obtaining y from x division, or the inverse of multiplication.

For every element of the set, there is an inverse element z such that $x+z = 0$. z is called the inverse of x in respect of the operation of addition. We write $z = -x$ and call the operation of obtaining z from x subtraction, or the inverse of addition.

† Sometimes called the unit element.

Clearly, for the set of natural numbers, there are no elements which possess an inverse in respect of the operation of addition; the only element which possesses an inverse in respect of the operation of multiplication is the element 1.

In general, consider any set S for which there is a neutral element e in respect of the operation $*$, and let x be any element of the set. Then if we can find an element y of the set such that

$$x * y = y * x = e,$$

we call y the inverse of x and denote it by x^{-1}. In symbols, $\forall x \in S, \exists x^{-1} \in S : x * x^{-1} = x^{-1} * x = e$.

Morphisms

For any set A, let $*$ denote a binary operation on A and let f be a function mapping A on B, $f : A = B$.

If x_1 and x_2 are any pair of members of A, we shall investigate whether there is any relation between $f(x_1)$, $f(x_2)$ and $f(x_1 * x_2)$.

Consider first the function $f(x) = x^2$, where x is a member of the set of real numbers $R^{\#}$, and the binary operation $*$ is multiplication; the set $R^{\#}$ is closed in respect of multiplication.

Then

$$f(x_1) = x_1^2, \quad f(x_2) = x_2^2, \quad \text{and} \quad x_1 * x_2 = x_1 x_2.$$

It follows that $f(x_1 * x_2) = (x_1 x_2)^2$ and hence

$$f(x_1) * f(x_2) = f(x_1 * x_2).$$

Next, consider the function $f(x) = e^x$, where $x \in R^{\#}$ and $*$ is the binary operation of addition. Then

$$f(x_1) = e^{x_1}, \quad f(x_2) = e^{x_2} \quad \text{and} \quad x_1 * x_2 = x_1 + x_2$$

It follows that

$$f(x_1 * x_2) = e^{x_1 + x_2},$$

and

$$f(x_1) * f(x_2) = e^{x_1} + e^{x_2}.$$

In this case there is no simple relation between $f(x_1 * x_2)$ and $f(x_1) * f(x_2)$. But, if we introduce a second binary operation of multiplication $\circ$ for the elements of set B, it is clear that

$$f(x_1) \circ f(x_2) = e^{x_1} \cdot e^{x_2} = e^{x_1 + x_2}$$

Thus

$$f(x_1) \circ f(x_2) = f(x_1 * x_2).$$

THE SYMBOLISM OF MODERN ALGEBRA

For any given set A, let f denote a function $f: A = B$ and let $*$ denote a closed binary operation on the set A. Then if we can find a binary operation $\circ$ on the set B such that, if x_1 and x_2 are any pair of elements of A,

$$f(x_1) \circ f(x_2) = f(x_1 * x_2),$$

we say that the function f is a morphism. We write

$$f:(A, *) = (f(A), \circ).$$

When the function is one–one, the morphism is called an isomorphism; when the function is many–one, the morphism is called a homomorphism.

Exercises 8(d)

1. Prove that the set of complex numbers $\{1, -1, i, -i\}$ is closed under the operation of multiplication.

2. $A = \{1, 2, 3, 4\}$. $\forall x, y \in A$, we define the operation $*$ as the remainder after dividing xy by 5. Prove that the set is closed for this operation and that the neutral element is 1. If $x \in A$, does x possess an inverse in respect of $*$?

3. S is the set of rational numbers, x and y are any members of the set and the operation $*$ is defined by $x * y = xy + x + y$. Prove that the set is associative and commutative in respect of the operation $*$ and that the neutral element of the set is 0. Check whether the elements of S have an inverse in respect of the operation $*$.

4. (i) For the operation of union of sets find the neutral element and state whether any set has an inverse.
   (ii) For the operation of intersection of sets find the neutral element and state whether any set has an inverse.

5. $\forall x, y \in R^\#$, the law of operation $*$ is defined by

$$x * y = x^2 y^2.$$

State whether the set is commutative, associative in respect of $*$.

6. Prove that the following functions are morphisms:

   (i) $f(x) = x^3$, $x \in R^\#$, $*$ represents the binary operation of multiplication.
   (ii) $f(x) = \log x$, $x \in A$, the set of positive real numbers, $*$ represents the binary operation of multiplication.
   (iii) $f(x) = x - 3$, $x \in R^\#$ and the binary operation $*$ on $R^\#$ is defined by $x * y = (x - 3)(y - 3) + 3$, $x, y \in R^\#$.

Groups

The concept of a group is one of the most important ideas in modern mathematics. A group is a set with the following properties in respect of a given operation $*$.

(i) The set is closed under the operation $*$.
(ii) The operation is associative.
(iii) The set possesses a neutral element.
(iv) Every element of the set possesses an inverse.

Expressing these results in symbolic form the set S is a group for the operation $*$ if

(i) $\forall x, y \in S, x * y \in S$.
(ii) $\forall x, y, z \in S, (x * y) * z = x * (y * z)$.
(iii) $\forall x \in S \; \exists e \in S: x * e = e * x = x$.
(iv) $\forall x \in S \; \exists x^{-1} \in S: x * x^{-1} = x^{-1} * x = e$.

We can combine conditions (iii) and (iv) in the single form:

$$\forall x, y, z \in S,$$

the relation $x * y = z$ has a unique solution when any two of x, y, z are given.

If, in addition to the four axioms, the commutative law

$$\forall x, y \in S, \qquad x * y = y * x,$$

is satisfied, we call the group a commutative group or an Abelian group.

Examples of groups

1. The set of real numbers $R^\#$ is a group in respect of the operation of addition, the neutral element being 0; the inverse of x is $(-x)$.

2. The set of real positive numbers $R^\#$ is a group in respect of the operation of multiplication, the neutral element being 1; the inverse of x is $(1/x)$.

3. The set of vectors in space is a group in respect of vector addition.

4. If n is a positive integer, the set

$$\{x \mid x = \exp(2\pi i t/n), t \text{ a natural number, } 1 \leq t \leq n\},$$

is a group for the operation of multiplication. For

(i) $\exp(2\pi i t/n) . \exp(2\pi i t_1/n) = \exp(2\pi i \overline{t+t_1}/n)$ or

$$\exp(2\pi i \overline{t+t_1-n}/n).$$

Hence the set is closed

(ii) $[\exp(2\pi i t/n . \exp(2\pi i t_1/n)] . \exp(2\pi i t_2/n) =$
$\exp(2\pi i t/n) . [\exp(2\pi i t_1/n) . \exp(2\pi i t_2/n)]$.

Hence the operation is associative.

(iii) $\exp(2\pi i n/n) = 1$ is a neutral element.

(iv) $\exp(2\pi it/n).\exp(\overline{2\pi in-t/n}) = \exp(2\pi in/n) = 1$. Hence, every element of the set possesses an inverse.

Klein's four-group

Consider the following operations applied to a point P in the xy-plane:
 (i) Leave P unchanged.
 (ii) Reflect P in OX.
 (iii) Reflect P in OY.
 (iv) Reflect P about O.

The first operation is called the identity operation and we shall denote it by I. We shall refer to the other operations as A, B, C, and define the operation $A * B$ as the combined effect of operation A followed by operation B, and so on for the other pairs of operations.

We can produce a structure table or Cayley table as follows:

*	I	A	B	C
I	I	A	B	C
A	A	I	C	B
B	B	C	I	A
C	C	B	A	I

The four transformations I, A, B, C form a group.

Any group of four elements with the above structure table is called a Klein four-group.

Order of a group

If a group has a finite number of elements, that number is called the *order* of the group.

If two groups have the same order and structure tables of the same form, the groups are said to be *isomorphic*.

Isomorphic groups

Given that
 (i) G is a group under the operation $*$;
 (ii) G_1 is a group under the operation $\circ$;
 (iii) there is a one–one relation between G and G_1, i.e. to each element of G there corresponds a unique element of G_1, and to each element of G_1 there corresponds a unique element of G;

(iv) if x, y are any two elements of G, and x_1, y_1 the corresponding elements of G_1, then to the element $x * y$ of G there corresponds the element $x_1 \circ y_1$ of G_1, we say that the groups G and G_1 are isomorphic.

It is left as an exercise for the reader to prove that
(a) the identity element of G corresponds with the identity element of G_1,
(b) to any element and its inverse in G there corresponds an element and its inverse in G_1.

Exercises 8(e)

1. State which of the following sets are groups:
   (i) the set of natural numbers under multiplication;
   (ii) the set of integers (positive, negative and zero) under addition;
   (iii) the set of real numbers under the operation $*$ defined by
   $$x * y = xy(x+y);$$
   (iv) the set of ordered pairs (x, y), where x and y are real, under the operation $*$ defined by
   $$(x, y) * (x_1, y_1) = (x+x_1, y+y_1);$$
   (v) The set $\{2^n \mid n \text{ an integer, positive, negative, or zero}\}$.

2. Prove that if x and y are elements of a set S which is a group under the operation $*$,
$$(x * y)^{-1} = y^{-1} * x^{-1}.$$

3. The residue of m (mod n) is defined as the remainder when m is divided by n, m and n being positive integers.
   Consider the set $S = \{1, 2, 3\}$ and the operation $*$ defined by
   $$x * y = \text{residue of } xy \text{ (mod 4)}.$$

We can construct a table, called a structure table, or a Cayley table for the behaviour of all pairs of elements of S under the operation $*$ in the form

$*$	1	2	3
1	1	2	3
2	2	0	2
3	3	2	1

Prove that S is not a group under the operation $*$.

4. Consider the set $S = \{1, 2\}$ and the operation $*$ defined by
$$x * y = \text{residue of } xy \text{ (mod 3)}.$$

Construct a structure table similar to that in Q(3) and prove that S is a group under $*$.

5. Consider the set $S = \{1, 2, 3, 4\}$ and the operation $*$ defined by

$$x * y = \text{residue of } xy \pmod 5.$$

Construct a structure table and prove that S is a group under $*$.

6. Let p be any prime number and consider the set

$$S = \{1, 2, 3, \ldots p-1\};$$

the operation $*$ is defined by

$$x * y = \text{residue of } xy \pmod p.$$

Prove that S is a group under $*$.

7. If S is the set $\{0, 1, 2, 3\}$ and the operation $*$ is defined by

$$x * y = \text{residue of } (x+y) \bmod 4,$$

prove that S is a group under $*$.

8. If S is a group for which $x^{-1} = x$ for all elements of the group, prove that the group is commutative.

9. Prove that in a Cayley table for a group S, each row (column) must contain all the elements of S.

If S is the group $S = \{a, b, c, d\}$ with identity element a and $b * b = d$, use the Cayley table to solve the equations $c * x = d$, $x * b = c$.

10. For a given operation $*$ and any element t, we define $t^0 = e$ (the identity element), $t^1 = t * e$, $t^2 = t * t$, $t^3 = t * t^2$, etc. $t^{-2} = t^{-1} * t^{-1}$, $t^{-3} = t^{-1} * t^{-2}$, etc.

Prove that the set $\{t^n \mid n \text{ an integer, positive, negative or zero}\}$ constitutes a group in respect of the operation $*$.

A group G with the property that there exists an element $t \in G$ such that every element of G is a power of t, is called a cyclic group, and t is called a generator of the group.

Prove that every subgroup of a cyclic group is cyclic.

11. State which, if any, of the following groups are cyclic under the operation of addition:
    (i) the group of rational integers;
    (ii) the group of rational numbers;
    (iii) the group of real numbers.

12. Prove that the set $R^\#$ is a group under the operation $*$ defined by
    (i) $\forall x, y \in R^\#$, $x * y = \sqrt[3]{(x^3+y^3)}$;
    (ii) $\forall x, y \in R^\#$, $x * y = \sqrt[5]{(x^5+y^5)}$.

13. Prove that the set $S = \{x \mid x$ is an integer, positive, negative or zero$\}$ is a group in respect of addition, denoted by $*$.

The set is mapped onto the set S_1 by the function f, where $f(x) = 2^x$. Prove that the set formed, $S_1 = \{2^x \mid x$ is an integer, positive, negative or zero$\}$ is a group in respect of the operation of multiplication, denoted by $\circ$. Prove that there is a one–one relation between the groups S and S_1, and that, if x and y are any two elements of S, $f(x*y) = f(x) \circ f(y)$. Hence establish that S and S_1 are isomorphic groups.

14. Prove that, if G is a commutative group and G_1 a non-commutative group, G and G_1 cannot be isomorphic.

15. Prove that the set of residues (mod n) form a group under addition, and that the set of the nth roots of unity form a group under multiplication.

Prove also that the two groups are isomorphic.

16. In a cyclic group of order n and generator t, prove that t^r is a generator if and only if r is prime to n.

17. (i) Prove that the set of real numbers x such that $0 < x \leq 1$ is not a group under the operation of multiplication.
(ii) Prove that the set $\{a+b\sqrt{3}\}$, where a and b are rational numbers and not both zero, form a group under multiplication. State whether they form a group under addition.

18. Prove that the set of complex numbers $\{x+i(0), x \in R^\#\}$ is isomorphic with the set of real numbers $\{x, x \in R^\#\}$ under the operation of (i) addition (ii) multiplication.

19. G is a group $\{x \mid x \in R^\#\}$ under the operation $*$ defined by $\forall x, y \in R^\#$, $x*y = x+y$.

G_1 is a group $\{x \mid x \in R^\#\}$ under the operation $\circ$ defined by $\forall x, y \in R^\#$, $x \circ y = \sqrt[3]{(x^3+y^3)}$.

Prove that G and G_1 are isomorphic.

20. $f(x) = 1/(x+3)$, where x is a member of the set of positive real numbers. Prove that f is a morphism in respect of the binary operation of addition, the induced binary operation $\circ$ being defined by
$$x \circ y = xy/(x+y-3xy).$$

21. The set $\{x_1, x_2, \ldots\}$, where the xs are measured values of a physical variable, is denoted by A. If $x \in A$, we denote by $e_x = f(x)$ the error in the measured value x, and denote by $r_x = \phi(x)$ the corresponding relative error.

Consider the relations between (i) $e_{x+y}, e_x,$ and e_y; (ii) $r_{xy}, r_x,$ and r_y. Prove that f is a morphism for the binary operation of addition and ϕ is a morphism for the binary operation of multiplication.

22. P and Q are subsets of two-dimensional space defined by
$$P = \{(x, y): x^2+y^2 = 1\},$$
$$Q = \{(x, y): x+y \leq 2\}.$$
State which, if any, of the following statements are true: (i) $P \cup Q = P$; (ii) $P \cap Q = \phi$; (iii) $P' \cap Q = P'$; (iv) $P' \subset Q'$; (v) $P' \cap Q' = P'$; (vi) $P \cap Q' = \phi$.

Answers to Exercises

Exercises 1(a) In the answers to Chapter 1, c stands for convergent and d for divergent.
1. c; d; c; c; c; d; d; c.
2. $x < 1$; all values; $x < 1$; all values except $x = 1$; all values; $x \leq 1$.

Exercises 1(b)
1. c; c; c; c.
2. c; c; c if $|x| < 1$, d if $|x| \geq 1$; c; c if $|t| < 1$; c; c if $-1 < y \leq 1$; c.
3. c; d; c if $|x| < \frac{1}{5}$; c if $|x| < 1$ and $|x| > 1$; d.

Exercises 1(c)
3. $R(z) > -\frac{1}{4}$; $2z+1$.
6. d.

Exercises 1(d)
2. $\dfrac{\pi}{2ab(a+b)}$; $\dfrac{1}{a^2-b^2}\log\dfrac{a}{b}$; $\dfrac{\pi}{2(a+b)}$.
3. $\dfrac{\pi^2}{2a^3}$; $\left(\dfrac{a}{\pi}, 0\right)$.
5. $\frac{1}{4}$; $\dfrac{\pi}{16}$; $\frac{1}{4}$; $\dfrac{3\pi}{16}$.
6. $\dfrac{\pi^2}{a^{11}}\left(\dfrac{63}{256}\right)$.
7. $\frac{1}{5}\log 2$; $\frac{1}{6}\log 2$; $\dfrac{1}{2n}\log 2$.
8. π; $\dfrac{3\pi}{8}$.
11. $\dfrac{4\pi}{9\sqrt{3}} - \dfrac{1}{3}$.
18. c; c.

Exercises 1(e)
4. $\dfrac{\pi}{\sqrt{a}}$.

Exercises 1(f)
4. I_1; I_3; I_2.
12. 0 in all cases.

Exercises 1(g)
2. 0·5772.

Exercises 1(h)

3. $\frac{\pi^2}{12}$ in all cases.

7. Note that $\frac{1}{(2n-1)^{\frac{1}{3}}} + \frac{1}{(2n+1)^{\frac{1}{3}}} - \frac{1}{n^{\frac{1}{3}}} = (2^{\frac{1}{3}}-1)\frac{1}{n^{\frac{1}{3}}} + O\left(\frac{1}{n^{\frac{7}{3}}}\right)$.

Miscellaneous problems on Chapter 1

2. d for all z except $z = 0$; c if $|z-1| < 1$, d if $|z-1| \geq 1$; c for all z.
4. c if $s < 0$; c; d.
5. 0.
7. $\frac{\pi}{32}; \frac{2\pi}{3\sqrt{3}}; \frac{1}{6} - \frac{1}{4}\log(\frac{9}{5})$.
8. c if $s > 1$.
9. c; c absolutely if $k > 1$; c if $0 < k \leq 1$, d if $k \leq 0$.
10. sum $= \log(1+x)$ if $|x| < 1$; sum $= \frac{1}{2}\log 8$ if $x = 1$; series diverges if $|x| > 1$ or $x = -1$.
12. $k > -1; \frac{2}{\sqrt{(1-k^2)}}\left[\frac{\pi}{2} - \tan^{-1}\frac{k}{\sqrt{(1-k^2)}}\right]$ if $-1 < k < 1$; $\frac{1}{\sqrt{(k^2-1)}}\left[\log\frac{k+\sqrt{(k^2-1)}}{k-\sqrt{(k^2-1)}}\right]$ if $k > 1$.
13. (i) c if $|x| < 1$, d if $|x| \geq 1$;
    (ii) c if $|x| > 1$, d if $|x| \leq 1$;
    (iii) c if $|x| < 1$, d if $|x| > 1$, c if $x = -1$, d if $x = 1$;
    (iv) c if $|x| < 1$, d if $|x| > 1$, c if $x = 1$, d if $x = -1$.
17. (i) result true; argument false.
    (ii) result true; argument false.
    (iii) result true; argument true.
    (iv) result false.
19. $e^z(1 + 7z + 6z^2 + z^3)$, yes.
20. $-1 \leq x < 1; \frac{1}{2}(1-\sqrt{5}) \leq x \leq \frac{1}{2}(1+\sqrt{5})$.
21. $\frac{1}{k\sqrt{(k^2-1)}}\log[k+\sqrt{(k^2-1)}]$.
22. c if $|t| = 1$.
23. d, d.
25. c, c, d, d.
28. d.

Exercises 2(a)

3.
$a^n \sin\left(ax + \frac{n\pi}{2}\right); \cos\left(x + \frac{n\pi}{2}\right); a^n \sin\left(ax + b + \frac{n\pi}{2}\right); a^n \cos\left(ax + b + \frac{n\pi}{2}\right)$.

ANSWERS TO EXERCISES

4. $2^{n-1}\cos\left(2x+\dfrac{n\pi}{2}\right)-2^{2n-1}\cos\left(4x+\dfrac{n\pi}{2}\right);$

$\tfrac{1}{2}5^n\cos\left(5x+\dfrac{n\pi}{2}\right)+\tfrac{1}{2}\cos\left(x+\dfrac{n\pi}{2}\right);$

$\tfrac{1}{2}5^n\sin\left(5x+\dfrac{n\pi}{2}\right)+\tfrac{1}{2}3^n\sin\left(3x+\dfrac{n\pi}{2}\right).$

5. $(-1)^{n-1}(n-1)!/x^n;\ (-1)^{n-1}(n-1)!a^n/(ax+b)^n;$

$(-1)^n n!\left[\dfrac{1}{(x+1)^{n+1}}-\dfrac{1}{(x+2)^{n+1}}\right];$

$(-1)^n n!\left[\dfrac{-3}{(x+3)^{n+1}}+\dfrac{5}{(x+5)^{n+1}}\right].$

Exercises 2(b)

1. $3^n\cos\left(3x+\dfrac{n\pi}{2}\right);\ 3^n\cos\dfrac{n\pi}{2};$

$4^n\sin\left(4x+\dfrac{n\pi}{2}\right);\ -4^n\sin\dfrac{n\pi}{2};$

$2^n e^x \sin\left(\sqrt{3}x+\dfrac{n\pi}{3}\right).$

3. $r^n e^{ax}\cos(bx+n\theta)$, where $r=\sqrt{(a^2+b^2)}$, $r\cos\theta=a$, $r\sin\theta=b$.

Exercises 2(c)

2. $e^x\{x^3+3nx^2+3n(n-1)x+n(n-1)(n-2)\};$

$x.2^n\cos\left(2x+\dfrac{n\pi}{2}\right)+n.2^{n-1}\cos\left(2x+\dfrac{n-1}{2}\pi\right).$

3. $n(n-1)(n-2)(n-3)\dfrac{6^4}{3^n}\sin\left(\dfrac{\pi}{6}+\dfrac{n\pi}{2}\right).$

4. $a=\dfrac{-3n}{p},\ b=\dfrac{3n(n+1)}{p^2},\ c=\dfrac{-n(n+1)(n+2)}{p^3}.$

9. $x^2\dfrac{d^{n+2}y}{dx^{n+2}}+2nx\dfrac{d^{n+1}y}{dx^{n+1}}+n(n-1)\dfrac{d^n y}{dx^n};\ n(n-1)\left[\dfrac{d^n y}{dx^n}\right]_{x=0}.$

10. $n(n-1)(n-2)(n-3)3^{n-4};\ n(n-1)(n-2)\left[\dfrac{d^{n-2}y}{dx^{n-2}}\right]_{x=0}.$

11. $f^{(n)}(0)+nf^{(n+1)}(0)=0.$

13. $x+\dfrac{x^3}{3}+\dfrac{2x^5}{15};\ x^3-\tfrac{1}{2}x^5;\ 1+x+\tfrac{1}{2}x^2-\tfrac{1}{8}x^4-\tfrac{1}{15}x^5.$

Exercises 2(d)

5. $f(x) = x - \dfrac{x^3}{3} + \dfrac{x^5}{6} - \ldots$

6. $0, \dfrac{1}{3n+1}, \dfrac{-1}{3n+2}$.

Miscellaneous problems on Chapter 2

2. $\dfrac{2^{2n-1}[(n-1)!]^2}{(2n)!}$; $k\dfrac{[(2n)!]^2}{2^{2n}(2n+1)![n!]^2}$.

5. $\dfrac{-4n(n-1)}{(n+1)(n+2)}$, $n \geq 4$.

7. $(-1)^{n-1}\dfrac{1}{2n-1}$, $n \geq 1$.

10. $\dfrac{(n-1)!}{2}\left[\dfrac{1}{(1-x)^n} + \dfrac{(-1)^{n-1}}{(1+x)^n}\right]$; diverges.

12. $\dfrac{\sin m\theta}{\cos \theta} = m \sin \theta - \dfrac{m(m^2-2^2)}{3!}\sin^3 \theta + \dfrac{m(m^2-2^2)(m^2-4^2)}{5!}\sin^5 \theta - \ldots$

Exercises 3(a)

2. applies to (i) and (iv).
3. 1·4, 2·5, 3·6.
5. Roots $x = 0$ and three roots in the interval (0, 1).
6. $x = 1$ (4-fold), $x = 2$ (3-fold) and 4 roots in the interval (1, 2).
7. $27b^2 + 4a^3 = 0$; $a^n(1-n)^{n-1} + n^n b^{n-1} = 0$.
8. $n(n-1)\left\{\dfrac{-b(n-1)}{2a(n-2)}\right\}^{n-2} + 2a = 0$;

$\left\{\dfrac{-b(n-1)}{2a(n-2)}\right\}^n - \dfrac{b^2}{a}\dfrac{(n-1)(n-3)}{4(n-2)^2} + c = 0$.

9. $a = -9$, $b = 4$, $c = 12$.
10. $x = -1$, $x = 4$ are double roots.

Exercises 3(b)

1. $(\tfrac{7}{3})^{\frac{1}{2}}$; $(\tfrac{27}{4})^{\frac{1}{3}}$.
2. $\tfrac{1}{3}[1 + \sqrt{13}]$.
3. e; $-\tfrac{1}{2}$; 0.
4. $1/\sqrt{2}$.

ANSWERS TO EXERCISES

Exercises 3(c)
1. $\frac{8}{3}$; 1; $-2e$; $\frac{1}{5}$.
2. $\frac{1}{12}$; $\frac{-2}{\pi}$; $\frac{1}{2}$; $\frac{\pi^2}{2}$.
3. $\frac{(0\cdot 2)^6}{6}$.
4. $\frac{(\frac{1}{2})^4}{4!} e^{\frac{1}{2}}$.

Exercises 3(d)
1. $a = 16$, $b = -1$.
2. $a = \frac{9}{4}$, $b = \frac{9}{4}$, $h = -\frac{7}{4}$.
3. $\frac{b^2}{a}, \frac{a^2}{b}$.
6. $22x - 6y - 9z + 14 = 0$.
8. The equation $\sin x = \sinh x$ has three solutions, $x = 0$, $\pm \alpha$. y is a maximum for $x = 0$ and a minimum for $x = \pm \alpha$.

Exercises 3(e)
1. 8; $0\cdot 405\ 45 < \log \frac{3}{2} < 0\cdot 405\ 49$.

Miscellaneous problems on Chapter 3
2. 1; -1; -2.
4. $\dfrac{2^{\frac{1}{2}n} \cos\left(\dfrac{n\pi}{4}\right)}{n!}$.
7. $e^{-\frac{1}{2}}$; 4.
10. Error $< 10^{-4}$.
14. $-\frac{1}{16}$; -1.

Exercises 4(a)
2. $\frac{1}{9}[6^n - (-3)^n]$.
4. $3 - n - (-2)^{n-1}$.
8. $u_n = \frac{1}{9}\{5 \cdot 2^n - 12(-1)^n(n+1) + 25(-1)^n\}$.

Exercises 4(b)
1. $9e - 24$.
2. $I_2 = (x^2 - 2)\sin x + 2x \cos x$; $I_3 = (x^3 - 6x)\sin x + (3x^2 - 6)\cos x$.
3. $\dfrac{181}{128} e^8 + \dfrac{3}{128}; \dfrac{\pi^3}{2} - 12\pi + 24; \dfrac{1}{\sqrt{2}}\left(\dfrac{-\pi^3}{64} + \dfrac{3\pi^2}{16} + \dfrac{3\pi}{2} - 6\right)$.
4. $1 + \dfrac{\pi}{4}$; $\frac{26}{15}$.
5. $I_1 = \frac{1}{2}\log(1 + \sqrt{2})$, $I_5 = \frac{1}{4}\sqrt{2} - \frac{1}{4}\log(1 + \sqrt{2})$.
$I_3 = \frac{1}{2}(\sqrt{2} - 1)$, $I_7 = \frac{1}{3} - \frac{1}{6}\sqrt{2}$.

242   MATHEMATICAL ANALYSIS AND TECHNIQUES

6. $(n-1)I_n = \sec^{n-2} x \tan x + (n-2)I_{n-2}$; $\frac{7}{8}\sqrt{2} + \frac{3}{8}\log(\sqrt{2}+1)$; $-\frac{464}{35}\sqrt{3}$.
8. $\frac{1}{32}(5e^4 - 1)$.

Exercises 4(c)

1. $\frac{3\pi}{32} - \frac{1}{4}$.
2. $naI_n + (2n-1)bI_{n-1} + (n-1)cI_{n-2} = x^{n-1}\sqrt{(ax^2+2bx+c)}$;
$\frac{1}{6}(4\sqrt{5} - 7\sqrt{2}) + \frac{1}{2}\log\left(\frac{2+\sqrt{5}}{1+\sqrt{2}}\right)$.
5. $u_4 = \frac{3\pi^2}{64} - \frac{1}{4}$; $u_5 = \frac{4\pi}{15} - \frac{149}{225}$.

Exercises 4(d)

1. s.c.; m.c.
2. $x = 6.5$ approximately.
3. 6·2, 2·57.
4. $\frac{2}{\sqrt{5}}$.

Exercises 4(e)

1. $y = x+1$, $y = -x-1$, $x = 1$.
3. $y+x-1 = 0$, $y-x-4 = 0$.
4. $(y-x+4)(2y+x)(x-3) = (3-4x-9y)$.
6. $y+2 = 0$, $x+y = 0$, $y-3x-4 = 0$.
7. $y = \pm x$.
9. $x = 1$; $y = \pm x$.

Exercises 4(f)

2. $r^3 = 2ap^2$.
3. $\frac{1}{2}a \sec\frac{\theta}{2} \tan\frac{\theta}{2} + \frac{1}{2}a \log\left(\sec\frac{\theta}{2} + \tan\frac{\theta}{2}\right)$.
5. $\frac{l}{p^2} = \frac{2}{r} + \frac{e^2-1}{l}$.
7. $a^2r^2 = p^2(25r^2 - 24a^2)$.
11. $r = \sin(\theta+k)$; $r^2 = \sin^3\left(\frac{2\theta}{3}+k\right)$.
12. $\frac{3\pi a}{2}$.
14. $r = a\sin(2\theta+k)$; $\frac{\pi a^2}{8}$.
15. $\frac{\sqrt{5}}{2}a + \frac{\log(2+\sqrt{5})}{4}a$.

Exercises 4(g)

1. $2\sqrt{\dfrac{r^3}{c}}$.

2. $a\dfrac{(\theta^2+1)^{\frac{3}{2}}}{\theta^2+2}$.

3. $r^3 = a^2p$; $\dfrac{a^2}{3r}$.

4. $\dfrac{a}{n+1}$.

Exercises 4(h)

1. $8x^3 = 27ay^2$.

2. $x^{\frac{2}{3}}+y^{\frac{2}{3}} = \left(\dfrac{2a}{3}\right)^{\frac{2}{3}}$.

Miscellaneous problems on Chapter 4

3. $y+x-2 = 0,\ y+2x+1 = 0,\ y-4x+3 = 0$.
4. 2.
7. $3a$.
8. $(r^2+1)^2 = 5p^2$.
9. $I_3 = \tfrac{7}{8}\pi a^5$.
10. $a\left\{\sqrt{2}-\dfrac{\sqrt{5}}{2}+\log\left(\dfrac{2+\sqrt{5}}{1+\sqrt{2}}\right)\right\}$.
11. $I_{m,n}\left(1-\dfrac{n^2}{m^2}\right)+\dfrac{n(n-1)}{m^2}I_{m,n-2} = \dfrac{\sin m\theta}{m}\cos^n\theta - \dfrac{n\cos m\theta}{m^2}\cos^{n-1}\theta\sin\theta$
12. $y+x-4 = 0,\ 2y-x = 0,\ y+3x+1 = 0$.
13. $x^{\frac{2}{3}}+y^{\frac{2}{3}} = \left(\dfrac{2c}{3}\right)^{\frac{2}{3}}$.
15. $\dfrac{a^n}{(n+1)r^{n-1}}$.
19. $x = -1$, area $= 3$.
21. $\tfrac{25}{27}a$; $\tfrac{2}{3}a$.

Exercises 5(a)

1. 7; 13.

3. $-\dfrac{1}{\sqrt{2}}$; -2.

4. $\tfrac{2}{5},\ \tfrac{4}{5},\ \tfrac{4}{5}$.
6. $-\tfrac{2}{3},\ -\tfrac{1}{3}$.
7. $\dfrac{\partial y}{\partial t} = -ab\sin(bt+cx)$.

9. $-6, 1, 2$.
10. $\frac{5}{9}, -\frac{4}{9}$.

Exercises 5(b)

2. (i) $(0, 0, 6)$ maximum; $(2, 0, 2)$ neither maximum nor minimum.
   (ii) $(0, 0, 0)$ neither maximum nor minimum.

5. $\dfrac{4\Delta^2}{a^2+b^2+c^2}$.

6. (i) $(0, 0, 0)$ maximum; $(2, 2, -16)$ neither maximum nor minimum.
   (ii) $(0, 0, 0)$ maximum; $(\frac{1}{3}, \frac{1}{3}, -\frac{2}{27})$ neither maximum nor minimum.

7. $(0, 0, 0)$ minimum; $(0, \pm 1, 2e^{-1})$ maximum; $(\pm 1, 0, e^{-1})$ neither maximum nor minimum.

Exercises 5(c)

3. $\dfrac{x^2}{a^2}+\dfrac{y^2}{b^2} = 1$.

7. $x^2+y^2 = c^2 \cos^2 \dfrac{\alpha}{2}$.

10. $2xy = k^2$.

13. $(ax)^{\frac{2}{3}} - (by)^{\frac{2}{3}} = (a^2+b^2)^{\frac{2}{3}}$.

Exercises 5(e)

1. $\dfrac{\partial z}{\partial x} = y(x^2+y^2)(5x^2+y^2)$; $\dfrac{\partial f}{\partial x} = yt^2 = y(x^2+y^2)^2$.

2. $44, 85, 6, 9$.

3. $\dfrac{\partial w}{\partial x} = 3x^2y+2y^2+2xy$; $\dfrac{\partial \phi}{\partial x} = 2xu = 2x^2y$.

4. $7, 18, -4, -18$.

Exercises 5(f)

1. $2 \cdot 6 \%$.

2. $x+3yt-5zt+t^2 = 0$.

3. 14%.

4. $S = 50, 716$; $\delta S = 935$ approximately.

5. $\delta A = \dfrac{0 \cdot 005}{2\sqrt{3}}$, $\delta B = \dfrac{0 \cdot 02}{2\sqrt{3}}$, $\delta C = -\dfrac{0.025}{2\sqrt{3}}$.
   $A = 60°5'$, $B = 60°20'$, $C = 59°35'$.

Exercises 5(g)

3. $\dfrac{d^2z}{dt^2} = \dfrac{\partial^2 f}{\partial x^2}\left(\dfrac{dx}{dt}\right)^2 + 2\dfrac{\partial^2 f}{\partial x \partial y}\left(\dfrac{dx}{dt}\right)\left(\dfrac{dy}{dt}\right) + \dfrac{\partial^2 f}{\partial y^2}\left(\dfrac{dy}{dt}\right)^2 + \dfrac{\partial f}{\partial x}\dfrac{d^2x}{dt^2} + \dfrac{\partial f}{\partial y}\dfrac{d^2y}{dt^2}$.

7. $\dfrac{\partial z}{\partial x} = \dfrac{\partial f}{\partial u}\dfrac{\partial u}{\partial x} + \dfrac{\partial f}{\partial v}\dfrac{\partial v}{\partial x} + \dfrac{\partial f}{\partial w}\dfrac{\partial w}{\partial x}$.

8. $2y^2\dfrac{\partial z}{\partial x}+(2x^2y-y^2z)\dfrac{\partial z}{\partial y} = yz^2-4xy-2x^2z$.

10. $(x^2+2xy)\dfrac{\partial z}{\partial x} = (2xy+y^2)\dfrac{\partial z}{\partial y}$.

Exercises 5(h)

2. (i) $\dfrac{\partial z}{\partial x}(xz+2y^2)+\dfrac{\partial z}{\partial y}(yz+2x^2) = 4xy-z^2$.

   (ii) $\dfrac{\partial z}{\partial x}(xz-x^2y)+\dfrac{\partial z}{\partial y}(xy^2-yz) = 0$.

   (iii) $(z-y)\dfrac{\partial z}{\partial x}+(x-z)\dfrac{\partial z}{\partial y} = (y-x)$.

3. $x+y+z-3a = 0$.
7. $-xy^2e^{xy}$.

13. $\dfrac{dv}{dx} = \dfrac{\begin{vmatrix} \dfrac{\partial f}{\partial x} & \dfrac{\partial f}{\partial y} & \dfrac{\partial f}{\partial z} \\ \dfrac{\partial g}{\partial x} & \dfrac{\partial g}{\partial y} & \dfrac{\partial g}{\partial z} \\ \dfrac{\partial h}{\partial x} & \dfrac{\partial h}{\partial y} & \dfrac{\partial h}{\partial z} \end{vmatrix}}{\begin{vmatrix} \dfrac{\partial g}{\partial y} & \dfrac{\partial g}{\partial z} \\ \dfrac{\partial h}{\partial y} & \dfrac{\partial h}{\partial z} \end{vmatrix}}$

Miscellaneous problems on Chapter 5

1. $27x^4 = 256y^3$.
5. $z = e^x \cosh kt - 1 + xt$.
8. 2.6%.
12. $(-\frac{4}{9}, \frac{8}{9}, \frac{32}{81})$ maximum; $(\frac{2}{3}, \frac{2}{3}, 0)$, $(-2, 2, 0)$, $(0, 0, 0)$ saddle-points.
16. $V = \left(r-\dfrac{a^2}{r}\right)\cos\theta$.
18. $\delta A = -\dfrac{11}{3000}$; $A \to 44°47'$ approximately.

Exercises 6(a)

2. $(0, 0, 0)$ minimum; $(\frac{1}{2}, \frac{1}{2}, e^{-1})$ saddle-point; $(-\frac{1}{2}, -\frac{1}{2}, e^{-1})$ saddle-point $(\frac{1}{2}, -\frac{1}{2}, 2e^{-1})$ maximum; $(-\frac{1}{2}, \frac{1}{2}, 2e^{-1})$ maximum.

6. $(0, 0, 0)$ saddle-point; $\left(0, \dfrac{1}{\sqrt{2}}, -\dfrac{1}{4}\right)$ minimum; $\left(0, -\dfrac{1}{\sqrt{2}}, -\dfrac{1}{4}\right)$ minimum greatest value 1 at $(\pm 1, 0, 1)$ on boundary; least value is $-\tfrac{1}{4}$.

8. $\left(0, \dfrac{2}{\sqrt{3}}, \dfrac{-16}{3\sqrt{3}}\right)$ minimum; $\left(0, \dfrac{-2}{\sqrt{3}}, \dfrac{16}{3\sqrt{3}}\right)$ maximum; $(2, 0, 0)$ saddle-point, least value -15, greatest value 15.

9. $(0, 0, 0)$ and $(\tfrac{4}{5}, -\tfrac{8}{5}, 0)$ are saddle-points; $(-\tfrac{2}{5}, -\tfrac{6}{5}, \tfrac{8}{5})$ maximum; $(\tfrac{2}{3}, -\tfrac{2}{3}, -\tfrac{8}{27})$ minimum, least value $-16 \cdot 9$; greatest value $29 \cdot 0$.

Exercises 6(b)

1. $6\sqrt{5}, 6\sqrt{7}$.
2. $\left(\dfrac{c}{p+q}\right)^{p+q} \left(\dfrac{p}{a}\right)^p \left(\dfrac{q}{b}\right)^q$.
3. $x = y = z = k$.

Exercises 6(c)

2. $\dfrac{48}{\sqrt{14}}$.

4. $\phi'(t) = \dfrac{1}{t+1}$ if $t > -1$; $\phi(t) = \log(t+1)$.

Miscellaneous problems on Chapter 6

3. $(t, 0, 0)$ saddle-point; $(0, 8, 0)$ saddle-point; $(-1, 4, -32)$ minimum.
4. $\dfrac{1}{\lambda}\left[\dfrac{\partial f}{\partial x}\mathbf{i} + \dfrac{\partial f}{\partial y}\mathbf{j} + \dfrac{\partial f}{\partial z}\mathbf{k}\right]$, where $\lambda^2 = \left(\dfrac{\partial f}{\partial x}\right)^2 + \left(\dfrac{\partial f}{\partial y}\right)^2 + \left(\dfrac{\partial f}{\partial z}\right)^2$.
6. $\dfrac{1}{\sqrt{3}}[\sqrt{(a^2+b^2+c^2)} - 3p]$.
7. (i) $\sqrt{5}, -1$; (ii) $5, 0$; (iii) $\tfrac{1}{2}, -\tfrac{1}{2}$; (iv) $\dfrac{2}{3\sqrt{3}}, \dfrac{-2}{3\sqrt{3}}$; (v) $4, \dfrac{-5}{6}$; (vi) $\dfrac{16}{25\sqrt{5}}, -\dfrac{16}{25\sqrt{5}}$.

Exercises 7(a)

1. $\tfrac{8}{15}$.
2. $12\pi a^2$.
4. $\tfrac{256}{15}a^2$.
7. $\dfrac{4a^2}{3}$.
8. $\dfrac{1}{8e^2}$.

Exercises 7(b)

1. -4π.
2. 0; $-\frac{2}{3}$.
3. -20.
4. $\frac{64}{\pi^2}[\pi-1-\sqrt{2}]$.
5. $\frac{337}{6}$.
6. $\frac{5}{12}a^2b^2$.
7. $-\frac{43}{105}a^3$.
8. 0.

Exercises 7(c)

1. $\frac{380}{3}$.
2. $\frac{225}{256}$.
3. $15a^3$.
4. $\frac{16}{21}$.
6. $\frac{1}{6}\pi ka^6$; $\left(\frac{16a}{7\pi}, 0\right)$.
7. $\frac{1}{18}$.
9. $\frac{\pi ab}{2}$.
11. $\frac{2}{105}$.
12. 4π.
13. $\frac{2}{3}a^2b \tan\theta$.
14. $\int\limits_{b/(a+b)}^{1} dy \int\limits_{0}^{b(1-y)/y} f(x,y)\,dx$; $\int\limits_{\frac{1}{2}}^{1} dy \int\limits_{1}^{\frac{1}{y}} f(x,y)\,dx + \int\limits_{0}^{\frac{1}{2}} dy \int\limits_{1}^{2} f(x,y)\,dx$.
15. $\frac{\pi}{8}+\frac{1}{30}$.
17. $-\frac{7\pi}{2}$.
18. (i) 252; (ii) 264; (iii) 274·5.
19. 0.
21. $-\frac{16}{3}$.

Exercises 7(d)

1. $a^2(\pi-2)$.

Exercises 7(e)
4. 93.
5. −1.
6. $12\pi - \dfrac{8}{5}$.

Exercises 7(f)
1. Negatively.
3. No, yes.

Exercises 7(g)
4. $\dfrac{\begin{vmatrix} \dfrac{\partial(f,\phi)}{\partial(x,u)} & \dfrac{\partial(f,\phi)}{\partial(x,v)} \\ \dfrac{\partial(f,\phi)}{\partial(y,u)} & \dfrac{\partial(f,\phi)}{\partial(y,v)} \end{vmatrix}}{\left\{\dfrac{\partial(f,\phi)}{\partial(u,v)}\right\}^2}$.

Exercises 7(h)
1. $\dfrac{14\pi}{3}; \dfrac{62\pi}{5}$.
3. $\dfrac{\pi\rho a^8}{32}$.
4. $\tfrac{1}{4}\tfrac{2}{5}$.

Miscellaneous problems on Chapter 7
1. 324.
2. $\dfrac{-19}{30}$.
5. $\dfrac{45}{32}\pi$.
6. $\tfrac{1}{32}(e+2e^4)$.
10. $\tfrac{1}{15}; \tfrac{1856}{15}$.
11. $\dfrac{5a^2}{2}$.
13. $e^{u^2+v^2}\left(\dfrac{4uv-2v}{4uv-1}\right)$.
14. $\tfrac{40}{3}$.
17. 36π.
21. $\dfrac{\pi ab(a^2+b^2)}{4}$.
23. $\dfrac{16a^3}{3}(\pi - \tfrac{4}{3}); 8a^2(\pi-2)$.

Exercises 8(a)
2. (i) $a = c$, $b = d$ or $a = d$, $b = c$. (ii) $a = c$, $b = d$.
3. 2^n, including ϕ.
5. (i) none (ii) A (iii) A, B, D, (iv) none.

Exercises 8(b)
1. (i).
5. $A' = \{x \mid x = 3t, t \in \varepsilon\}$.

Exercises 8(d)
2. Yes. Inverse of x is $-x/(x+1)$.
4. (i) ϕ; ϕ has inverse but no other set.
   (ii) ε; ε has inverse but no other set.
5. Yes, no.

Exercises 8(e)
1. (ii), (iv), (v).
9. $x = c$, $x = d$.
22. (vi).

Index

Asymptotes, 94
Area of closed curve, 175

Binomial Theorem, 70

Catenary, 106
Contact of curves, 77
Convergence of series, 1–39
  Absolute convergence, 10
  Alternating positive and negative terms, 9
  Cauchy's test, 5
  Circle of convergence, 13
  Complex terms, 11
  D'Alembert's test, 4
  Derangement, 32
  Differentiation, 41
  Integral test, 28
  Kummer's test, 6
  Uniform convergence, 13
Convergence of integrals, 15, 28
Curvature, 76
  Intrinsic equation, 105
  Polar coordinates, 101
  Tangential polar coordinates, 103
Curves
  Area of closed curve, 175
  Cycloid, 106
  Cycloidal pendulum, 111
  Envelopes, 128
  Epicycloid, 133
  Evolute, 131
  Length in polar coordinates, 98

D'Alembert's test for convergence, 4
Difference equations, 87
Directional derivatives, 166
Double Integrals
  Conditions for existence, 191, 196
  Change of variable in, 206, 212
  Definition, 190

Envelopes, 128
Epicycloid, 133
Euler's constant, 30
Evolute, 131
Experimental errors, 138

Functions, 225
  Codomain, 225
  Domain, 225
  Functions of more than one variable, 118
  Homogeneous functions, 168
  Maximum and minimum values of, 124, 157, 162

Green's theorem, 202
Groups, 231
  Isomorphic, 233
  Klein's four group, 233
  Order of group, 233

Integrals
  Cauchy's Principal value, 20
  Convergence of, 15, 19, 28
  With infinite integrands, 19
  With parameter, 168
Intrinsic equation of curve, 105

Jacobians, 209

Kummer's test for convergence of series, 6

Lagrange
  Method of undetermined multipliers, 164
  Remainder in Maclaurin's theorem, 69, 79
Laplace differential equation, 148
Leibnitz theorem, 47
L'Hôpital's rules for limits, 73
Line integrals, 187
Line of best fit, 125

Mapping, 206, 225
Maclaurin's theorem, 50, 67
  Remainder after n terms: Cauchy's form, 69, 79; Lagrange's form, 69, 79
Maximum and minimum values of functions of one variable, 72
Maximum and minimum values of functions of more than one variable, 124, 157, 162, 164
Mean value theorems, 63–67

Morphisms, 230
Multiple roots of equations, 60
Normals to surfaces, 139

Partial differentiation, 119, 120
Planimeter, 188
Potential of field, 121
Power series, 40–57
   Circle of convergence, 13
   Differentiation of, 41
   Integration of, 43
   Standard functions, 68

Reduction formulae for integrals, 87
Regions, simply connected, multiply connected, 91
Rolle's theorem, 58

Sets, 222
   Commutative laws, 227
   Intersection of sets, 223
   Empty sets, 223
   Intersection of sets, 223
   Neutral elements, 229
   Subsets, 223
   Union of sets, 223
   Universal sets, 224
   Disjoint sets, 223
Simpson's rule, 81
Small changes in connected variables, 135
Stability of mechanical systems, 111
Stationary values, 72, 73
Surface integrals, 200

Tangential polar equation of curve, 98
Taylor's theorem, 67, 76, 72
   For functions of two variables, 157

Volumes of solids, 197